普通高等教育"十三五"规划教材

特种加工

(双语教材)

夏法锋 马春阳 王金东 赵海洋 曹梦雨 ◎ 主编
国 雪 ◎ 审核

NON-TRADITIONAL MACHINING

内 容 提 要

本书以国内外前沿的特种加工相关理论和工艺为基础，系统地介绍了特种加工的先进制造工艺，主要包括加工原理、加工过程、加工特点、加工设备及其应用等，并涉及电化学、声学、光学、电磁学、材料物理学等领域的知识；重点介绍了电火花加工、电火花线切割加工、电化学加工、激光加工、超声波加工等相关知识。

本书具有体系全、覆盖面广、理论与实践并重等特点，既可作为高等院校本科生、函授生的教材，也可作为企业工程师的参考书。

图书在版编目（CIP）数据

特种加工：英汉对照 / 夏法锋等主编．—北京：中国石化出版社，2020.6
双语教材
ISBN 978-7-5114-5799-8

Ⅰ.①特… Ⅱ.①夏… Ⅲ.①特种加工-双语教学-高等学校-教材-英、汉 Ⅳ.① TG66

中国版本图书馆 CIP 数据核字（2020）第 075238 号

未经本社书面授权，本书任何部分不得被复制、抄袭，或者以任何形式或任何方式传播。版权所有，侵权必究。

中国石化出版社出版发行
地址：北京市东城区安定门外大街 58 号
邮编：100011　电话：(010)57512500
发行部电话：(010)57512575
http://www.sinopec-press.com
E-mail: press@sinopec.com
北京富泰印刷有限责任公司印刷
全国各地新华书店经销

*

787×1092 毫米 16 开本 10 印张 191 千字
2020 年 6 月第 1 版　2020 年 6 月第 1 次印刷
定价：38.00 元

前　言

20世纪40年代发明的电火花加工，开创了用软工具和不靠机械力来加工硬工件的特种加工方法。50年代以后先后出现了电子束加工、等离子弧加工和激光加工，这些加工方法不用成型的工具，而是利用密度很高的能量束流进行加工。对于高硬度材料和复杂形状、精密微细的特殊零件，特种加工有很大的适用性和发展潜力，在模具、量具、刀具、仪器仪表、飞机、航天器和微电子元器件等机械制造中也得到越来越广泛的应用。

特种加工的发展方向主要是：提高加工精度和表面质量；提高生产率和自动化程度；发展几种方法联合使用的复合加工；发展纳米级的超精密加工等。由于激烈的市场竞争，产品更新换代日益加快，同时产品要求具有很高的强度质量比和性能价格比，且正朝着高速度、高精度、高可靠性、耐腐蚀、高温高压、大功率、尺寸大小两极分化的方向发展。各种新材料、新结构、形状复杂的精密机械零件大量涌现，对机械制造业提出了一系列迫切需要解决的新问题。

近年来，随着电子技术、计算机技术和自动控制技术的发展，特种加工技术获得了飞跃发展和广泛应用。为适应特种加工技术的发展，全国众多工科院校纷纷开设了特种加工技术的必修或选修课程。本书根据工程应用型本科机械专业人才培养特点，用中文和英文进行编写。本书作为机械类特种加工教材，用中文和英文全面介绍了特种加工概述、电火花成型加工技术、电火花线切割加工技术、电化学加工、快速成型加工技术、激光加工技术、等离子体加工技术等知识。

本书由夏法锋教授、王金东教授、马春阳副教授、赵海洋副教授、曹梦雨讲师主编（排名不分先后），具体分工如下：第 1 章和第 2 章由夏法锋教授编著（约 5 万字），第 2 章、第 7 章和第 8 章由马春阳副教授编著（约 8 万字），第 3 章、第 4 章由王金东教授编著（约 4 万字），第 5 章由赵海洋副教授编著（约 1 万字），第 6 章由曹梦雨讲师编著（约 1 万字）。在编写过程中，本书得到了东北石油大学多位教师的鼎力协助，另外，本书编写过程也参考了一些文献，在此表示衷心感谢！

特种加工技术是一门不断发展的综合性交叉学科，涉及知识面广、学科多，由于编者水平和经验有限，疏漏在所难免，恳请读者批评指正。

目　　录

前言

Unit 1　Introduction　概论　001

1.1　Definition of Non-Traditional Machining　特种加工的定义 ······················001

1.2　Property of Non-Traditional Machining　特种加工的特点 ······················002

1.3　The machining revolution　加工方式的革命 ······················003

1.4　The two phases of the computer revolution　计算机革命的两个阶段 ······················005

1.5　Six Basic Processes to Alter Material in Non-Traditional Machining
　　　特种加工过程中六种去除材料的基本手段 ······················006

1.6　Machining Speed and Accuracy　加工速度和精度 ······················009

1.7　The Future　未来发展趋势 ······················009

　　　Questions　习题 ······················011

Unit 2　Electrical discharge machining　电火花加工　012

2.1　Fundamentals of electrical discharge machining　电火花加工的基础 ······················012

2.2　How does EDM work　电火花加工机理 ······················015

2.3　Dielectric Oil and Flushing for EDM
　　　在电火花加工过程中的工作液及其冲洗作用 ······················020

2.4　EDM Applications　电火花加工的应用 ······················032

Ⅱ | Non-Traditional Machining
特种加工

    Questions　习题 ·· 034

Unit 3　Wire Electrical Discharge Machining　电火花线切割加工　035

    3.1　Fundamentals of Wire Electrical Discharge Machining
        电火花线切割加工基础 ··· 035
    3.2　How does Wire EDM work　WEDM 的工作过程 ··· 038
    3.3　Proper Procedures　WEDM 的加工步骤 ·· 043
    3.4　Advantages of Wire EDM for Die Making
        WEDM 在模具制造方面的优势 ·· 047
    3.5　Wire EDM Applications　WEDM 的应用 ·· 052
    Questions　习题 ·· 054

Unit 4　Electrochemical Machining　电化学加工　056

    4.1　Fundamentals of ECM　电化学加工基础 ·· 056
    4.2　How does Electrochemical Machining work　电化学加工过程 ··················· 064
    4.3　Advantages of Electrochemical Machining　电化学加工的优点 ·················· 067
    4.4　Disadvantages of Electrochemical Machining　电化学加工的缺点 ············· 070
    4.5　Pulse Electrochemical Machining　脉冲电化学加工 ··································· 072
    4.6　ECM Applications　电化学加工的应用 ·· 077
    Questions　习题 ·· 080

Unit 5　Lasers Machining　激光加工　081

    5.1　Fundamentals of Lasers　激光加工基础 ··· 081
    5.2　Laser Component　激光加工器的组成 ·· 087
    5.3　Various Lasers and Their Configurations　各种激光器及其配置 ················ 093
    5.4　Traveling Methods of Laser Cutting Machines　激光切割器的移动方式 · 099
    5.5　Lasers Machining Applications　激光加工的应用 ······································· 100
    Questions　习题 ·· 108

Unit 6　Photochemical Machining　光化学加工　109

 6.1　Fundamentals of Photochemical Machining　光化学加工的基础 …………109

 6.2　Advantages of Photochemical Machining　光化学加工的优点 ……………115

 6.3　Disadvantages of Photochemical Machining　光化学加工的缺点…………116

 6.4　Applications of Photochemical Machining　光化学加工的应用 ……………117

 Questions　习题 ………………………………………………………………………118

Unit 7　Ultrasonic Machining　超声波加工　119

 7.1　Fundamentals of Ultrasonic Machining　超声波加工原理 …………………119

 7.2　Ultrasonic Machine　超声波加工机 ……………………………………………123

 7.3　Applications of Ultrasonic Machining　超声波加工的应用 …………………130

Unit 8　Others non-traditional Machining　其他种类的特种加工　132

 8.1　Plasma Cutting　等离子加工 ……………………………………………………132

 8.2　Water jet and Abrasive Waterjet Machining　水射流和磨粒流加工 ………138

 8.3　Electron Beam Machining　电子束加工 ………………………………………145

 8.4　Ion Beam Machining　离子束加工 ……………………………………………146

 Questions　习题 ………………………………………………………………………147

References ……………………………………………………………………………148

附录　Words and Expressions ……………………………………………………149

Introduction 概论   1

1.1 Definition of Non-Traditional Machining
特种加工的定义

When people hear the word "machining", they generally think of machines that utilize mechanical energy to remove material from the workpiece. Milling machines, saws and lathes are some of the most common machines using mechanical energy to remove material. All traditional forms of metal cutting use shear as the primary method of material removal. However, Non-traditional machining is a machining method, which uses non-mechanical energy (such as electricity, chemical, light, sound, heat, etc.) to remove unwanted material from a workpiece.

In today's highly competitive world, it is essential to understand the non-traditional machining processes. Because many non-

当人们听到"加工"这个词语的时候，通常会想到那些利用机械能从工件上去除材料的机器。研磨机、锯、机床这些使用机械能去除工件材料的机器便是其中最普通的一部分。所有的传统加工方式都是利用对工件材料的切割作为最基本的材料去除方式。然而，特种加工却是使用其他种类的能量（例如电能、化学能、光能、声能、热能等）来去除材料、加工工件。

在当今这个竞争激烈的社会，了解特种加工的过程是十分必要的。因为很多种类的特种加

traditional processes have major advances, every manufacturer needs to learn and understand the advantages of these latest technologies. For example, wire EDM has increased its speed of cutting up to ten times faster from when it was introduced. Likewise the capacity in workpiece weight and height has also increased.

Non-traditional machining is different from the traditional machining. Due to the rapid advances of technology, many traditional ways of today's machining are performed with the so-called "non-traditional" machines. Manufacturers are realizing results in achieving excellent finishes, high accuracies, cost reductions, and much shorter delivery times.

The purpose of this book is to educate engineers, designers, business owners, and those making machining decisions to understand and to be able to use the many non-traditional machining methods, and thus make their companies more profitable.

工都具有其重要的先进性，而每一个制造商都需要学习掌握最新技术的众多优点，例如，现在的电火花线切割加工的速度已经比它刚被发明出来时快10倍。其能加工工件的质量和尺寸也都上了一个台阶。

特种加工和传统加工不同。随着科技的不断进步，许多传统的加工方式都使用了所谓的"特种"机床。制造商们已经认识到了特种加工能获得较好的表面加工质量、较高的加工精度，并且能减少成本、缩短加工时间。

本书编写的目的在于使工程师、设计师、企业经营者和决策者了解并能够运用各种特种加工方法，从而使其所在的公司获得更多的盈利。

1.2 Property of Non-Traditional Machining
特种加工的特点

（1）It does not mainly rely on mechanical energy, but the use of other forms of energy

（1）特种加工并不主要依赖机械能去除工件材料，而是利用

(such as electricity, chemical, light, sound, heat, etc.) to remove the workpiece material.

(2) The hardness of the tool can be less than that of the workpiece material.

(3) There is no significant mechanical cutting force between the tool and workpiece during the machining process.

其他形式的能量（例如电能、化学能、光能、声能、热能等）来去除工件材料。

（2）刀具的硬度可以小于工件材料的硬度。

（3）在加工过程中，刀具和工件之间不存在明显的切削力。

1.3 The machining revolution　加工方式的革命

From the earliest ages, individuals learned to use various hand tools to shape objects. As knowledge increased, the use of tools also increased. The industrial revolution arrived and introduced more sophisticated and precise tools such as drill presses, lathes, and milling machines.

Another revolution came—CAD/CAM (computer aided design/computer aided machining). Instead of manually moving machines, computerized programs were downloaded into machines and the operations proceeded automatically. The use of these machines dramatically increased productivity.

With the addition of high speed computers, these machines achieved faster processing times. Then fuzzy logic was introduced. Unlike bi-level logic, which

很久以前，人们就学会了用各种各样的手工工具来使物体成型。随着科技的进步，人们对工具的应用也越来越广泛。工业革命的降临也带来了许多更复杂、更精密的工具，例如钻孔机、精压机、机床和研磨机等。

另一场革命接踵而来——CAD（计算机辅助设计）和CAM（计算机辅助制造）。与传统手动控制机器的方式不同，数控程序被下载到机床中，并且其运行过程是完全自动的。此类机器的应用显著地提高了生产率。

随着高速计算机的加入，数控机床获得了更快的加工速度，模糊逻辑学也应运而生。与精密逻辑学不同，模糊逻辑学的语句

states that a statement is either true or false, fuzzy logic allows a statement to be partially true or false. Machines equipped with fuzzy logic "think" and respond quickly to minute variances in machining conditions. They can then lower or increase power settings according to messages received.

Another great innovation was virtual reality for prototyping and manufacturing. Virtual reality is an artificial environment where any of the human senses such as sight, sound, taste, touch, or smell can be simulated. Virtual reality is simply a computerized experience of real world situations.

In virtual prototyping and manufacturing, a three-dimensional CAD image can be made into a solid part or model directly from the computer. In virtual reality, engineers can visualize and manipulate a 3D model and make any necessary changes before incurring hard tooling costs. This process of rapid prototyping dramatically reduces time and costs.

Other innovations include automatic tool changers, robots, workpiece and pallet changers, and artificial intelligence that enable machines to "think" through complex machining sequences.

表达对错均可，模糊逻辑学允许一个语句有一部分是错的。装载有模糊逻辑学程序的机器对于加工情况实时的变化能做出迅速的"思考"和反应。依据处理器接到的信息，可以减少或增加能源供应量以满足加工要求。

另一个巨大的进步就是对于样机研究和加工的虚拟现实。虚拟现实是一个人造的环境，在这里，人类的任何感官例如视觉、听觉、味觉、触觉或嗅觉都可以被模拟出来。虚拟现实只是对真实情况的电脑模拟结果。

在虚拟造型和制造中，可以使用三维计算机辅助设计技术进行三维实体建模。在虚拟现实中，设计者可以想象并制造一个3D模型，并且在真实加工之前做出一些必要的调整。这个快速造型的过程显著地减少了加工时间和费用。

其他的技术革新包括：自动化工具的改进、机器人的出现、工件和操作面板的改良以及使机器能识别复杂加工序列的人工智能。

1.4 The two phases of the computer revolution 计算机革命的两个阶段

(1) Computer-controlled machines, such as lathes, mills, and machining centers, using traditional machining methods which use hard tooling.

(2) Computer-controlled machines using non-traditional machining methods which use processes other than hard tooling like EDM, wire EDM, abrasive flow and ultrasonic machining, photochemical and electrochemical machining, plasma and precision plasma cutting, water-jet and abrasive water-jet machining, laser beam machining, and so on. See Figure 1-1.

（1）如车床，磨床和加工中心这些数控机床，都使用了利用传统的硬质刀具加工的方法。

（2）利用特种加工手段的数控机床使用的是电火花加工，电火花线切割加工，超声波加工，光化学加工，电化学加工，等离子及精确等离子体加工，水射流、磨料流加工，激光加工以及快速成型技术，如图1-1所示。

(a)EDM
电火花加工

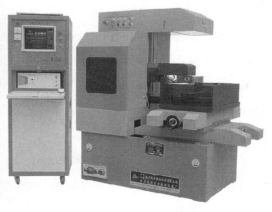

(b)Wire EDM
电火花线切割加工

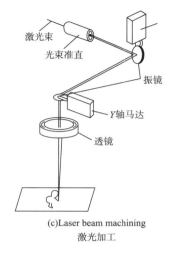

(c)Laser beam machining
激光加工

(d)Water-jet machining
水射流加工

Figure 1-1　Some non-traditional machining methods
部分特种加工手段

1.5　Six Basic Processes to Alter Material in Non-Traditional Machining　特种加工过程中六种去除材料的基本手段

There are basically six processes used in non-traditional machining to alter material: electricity, water, abrasives, chemicals, plasma, and light. Understanding these processes will help decision makers to evaluate the opportunities and limitations of these operations.

1.5.1　Electricity

Wire EDM and EDM use electricity to cut electrically conductive material by means of spark erosion.

1.5.2　Water

Water-jet machining uses high pressure

特种加工有六种去除材料的基本手段：电能加工、水射流加工、磨料流加工、化学加工、等离子加工和光能加工。了解这些手段对决策者评价这些加工手段的优点与缺陷是很有帮助的。

1.5.1　电能加工

电火花线切割加工和电火花加工都可以通过放电腐蚀的方式使用电能切割导电工件。

1.5.2　水射流加工

水射流加工使用高压水射流

water (up to three times the speed of sound) to cut material. Since water is used, any material can be cut if it is soft enough for high pressure water to penetrate.

1.5.3 Abrasives

Abrasive water-jet adds garnet, a silicate abrasive material, to a high pressure water jet. This allows any material to be cut. Abrasive flow machining is used to deburr and polish parts by forcing a semisolid abrasive media through the workpiece. Ultrasonic machining uses fine, water-based abrasive slurry to machine parts. The vibrating machine causes abrasives to remove material.

1.5.4 Chemicals

Photochemical machining relies on chemicals to remove exposed material to etch or cut parts. Electrochemical machining combines chemicals and electricity to remove material through a deplating process. As a salt solution electrolyte surrounds an electrode, an electrical current passes from the electrode to the workpiece removing the material.

1.5.5 Plasma

Plasma and precision plasma cutting systems utilizes ionized gas to cut electrically conductive materials.

（其速度可达到三倍声速）来切割工件材料。

1.5.3 磨料流加工

磨料流把石榴石（一种硅酸盐磨粒材料）加入高压水射流之中，因此它可以切割任意种类的材料。磨料流加工一般通过将高压半固体磨料流喷射到工件表面来达到去毛刺和抛光工件的目的。超声波加工使用极细的水基磨料浆来加工工件，机器传导给磨料的震动使磨料获得足够的能量来移除材料。

1.5.4 化学加工

光化学加工依靠化学物质来去除暴露在外的材料，从而达到蚀刻或者切割工件的目的。电化学加工综合使用了化学能和电能，通过去除镀层的方式来移除材料，当电极被电解液包围的时候，从电极反射的脉冲电流移除工件材料。

1.5.5 等离子加工

等离子加工和精细等离子切割系统使用电离了的气体来切割导电材料。

1.5.6 Light

Lasers rely on highly magnified light to alter materials. Since light is used, both electrical and non-electrical material can be processed. Besides cutting, lasers are used for welding, cladding, alloying, and heat treating.

The six processes to alter material in non-traditional machining are electricity, water, abrasives, chemicals, ionized gas, and light. See Table 1-1.

1.5.6 光能加工

激光加工使用高度放大的光来移除材料。由于使用的是光能,不管材料导不导电,激光加工都可以对其进行加工。除了激光切割之外,激光加工还用于焊接、激光镀、铸造合金、热处理、打标识、激光钻孔。

特种加工的六种移除材料的媒介分别是:电能、水射流、磨料流、化学能、电解液和光,如表1-1所示。

Table 1-1　Six processes in non-traditional machining　特种加工的六种加工手段

Item 加工媒介	NTM method 特种加工手段
Electricity 电能	EDM and WEDM 电火花加工和电火花线切割加工
Water 水	Water-jet Machining 水射流加工
Abrasives 磨料流	Abrasives Water-jet, Abrasive Flow, Ultrasonic Machining 磨料水射流加工,磨料流加工,超声波加工
Chemicals 化学物质	Electrochemical Machining, Photochemical Machining 电化学加工,光化学加工
Ionized Gas 电解液	Plasma Cutting 等离子加工
Light 光	Laser Beam Machining 激光加工和快速成型加工*

注:*一些快速成型系统并不使用激光;一些热硬性塑料和其他系统使用曝光时变硬的光敏聚合物。

1.6　Machining Speed and Accuracy　加工速度和精度

The processing speed and the machining precision of the above-mentioned processing method are very different. Understanding the different characteristics of different systems can greatly affect the number of manufacturers profit. For example: metal foil can be laser cutting the way quickly cut, but the laser processing can only cut a workpiece. The use of EDM, metal sheet pieces can be stacked together a few pieces of processing, and the workpiece processing more free, more economical and more accurate.

上述加工方法的加工速度和加工精度差别很大。了解不同系统的不同特性可以很大程度地影响厂商的利润。例如：金属薄片可以用激光加工的方式很快切割完毕，但是激光加工每次只能切割一个工件。而使用电火花加工，金属薄板件可以几块叠在一起加工，而且工件的加工更自由、经济、精确。

1.7　The Future　未来发展趋势

Companies can only remain successful if they actively endeavor to keep their operations competitive. The moment they become careless, others will arise and work more competitively. Eventually, due to their higher machining costs, these inefficient companies will be put out of business. The benefit of the free enterprise system is that surviving companies are the ones that produce the best products at the lowest prices.

Therefore, to remain successful,

只有通过积极努力，保证自己的运作更加具有竞争力，一个企业才可能保持成功，否则，会立马被其他企业赶超。最终，加工成本过高，效率最低的企业将会被行业淘汰。这种自由竞争模式的好处就是最终存活下来的企业一定是以最低成本生产出最优质产品的企业。

因此，为了保持成功，企业

companies need to keep informed of the newest technologies in order to remain competitive. In addition, they need to train their employees to work efficiently and accurately.

Non-traditional machining has great effect on the traditional manufacturing industry, and the effects are listed as follows:

(1) It has improved the machinability of the material. Appropriate non-traditional machining not only can process hard alloy, stainless steel, and hardened steel, but also can ceramic and glass.

(2) It has changed the typical process route of parts.

(3) It shortens the cycle of new products.

(4) It affects the structural design of parts.

(5) It reconsiders the traditional structure and processing technology.

(6) It becomes the primary means of the micro/nano fabrication.

The future belongs to those who remain at the forefront of the latest technology and apply the latest technology to remain competitive. However, the companies those tend to be self-satisfied and refuse to use new technologies will lose competitiveness.

必须对高新技术了如指掌，以保证其在行业中的竞争力。另外，他们需要培训雇员，以使其能更加高效、更加精确地工作。

特种加工对传统加工产生的影响是巨大的，大致影响如下：

（1）提高了材料的可加工性，适当的特种加工手段不仅可以加工硬质合金、不锈钢、硬钢，也可以加工陶瓷和玻璃；

（2）改变了零件加工的传统工艺路线；

（3）缩短了新产品的试制周期；

（4）影响了零件的结构设计；

（5）需要重新审视传统的结构工艺性；

（6）成为微细加工/纳米技术的主要手段。

未来属于那些站在最新技术前沿并应用最新技术保持竞争力的人。那些停滞不前、拒绝使用新加工技术的企业将会失去竞争力。

毫无疑问，任何读到本章的读者都会对学习新的特种加工感兴趣。对于这样的读者，该书考察了现在市场上用得到的、十分振奋人心的手段。因此，他们可以做出更高效、更有竞争力的抉择。

Questions　习题

1. List three items that have revolutionized machining since the industrial revolution.
2. Identify and describe the two phases of the computer revolution.
3. What are the characteristics of Non-Traditional Machining?
4. List the six processes and the machines that are used in non-traditional machining.

1. 列举三种工业革命时期使机器进化的项目。
2. 识别并描述计算机革命的两个现象。
3. 特种加工的特点是什么？
4. 列举特种加工的六种加工手段以及其使用的机器。

2 Electrical discharge machining
电火花加工

2.1 Fundamentals of electrical discharge machining
电火花加工的基础

Electrical discharge machining (EDM) Witch is known as ram EDM, sinker EDM, die sinker, vertical EDM, and plunge EDM, as shown in Figure 2-1, is generally used to produce blind cavities.

When it is necessary to process a blind hole, it is essential to process a desired shape of the electrode. The current is then released through the electrode, and the shape is regenerated on the workpiece by the surrounding of the insulating liquid, as shown in Figure 2-2.

Discharge is a form of EDM that produces effects similar to lightning strikes our earth. Similarly, when a screwdriver short-circuits the car body and battery, you

电火花加工，即所谓的 ram EDM、sinker EDM、die sinker、vertical EDM 和 plunge EDM，图 2-1 是通常用来加工盲孔的电火花加工机床。

当需要加工一个盲孔时，则需要加工一个所要求形状的电极。然后通过电极释放电流，在绝缘液体的环绕下，在工件上就重新生成了它的形状，如图 2-2 所示。

放电是电火花加工的一种形式，它产生的效果可以看作类似闪电袭击地球。同样，当一个螺丝刀短路了汽车车身和电池时，

Figure 2-1　Electrical discharge machining
电火花加工机床

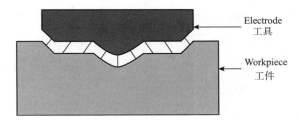

Figure 2-2　EDM process uses a formed electrode to remove material
电火花加工过程利用一个成型的电极来去除材料

can see how the current removes the metal material.

In 1889, Benjamin Chew Tilghman, of Philadelphia, received a U.S. patent (Patent No. 416,873) entitled, "Cutting Metal By Electricity". Although Tilghman had developed the concept of electrical discharge machining, spark erosion devices between World War I and World War II were used

就可以目睹电流是如何去除金属材料的。

1889 年，来自美国费城的 Benjamin Chew Tilghman 获得了一项美国专利（专利号：416,873），即"通过电流来切削金属"。尽管 Tilghman 早已发展了电火花加工的概念，但是在一战和二战期间火花侵蚀装置主要还是被用来

primarily to remove broken drills and taps. These early machines were very inefficient and difficult to use.

Then two Russian scientists, Boris R. and Natalie I. Lazarenko (husband and wife) made two important improvements. First they developed the R-C relaxation circuit which provided a consistent pulse control. Second they developed a servo control unit which maintained a consistent gap allowing efficient electrical discharges.

These two developments made EDM a more dependable means of production. However, the process still had its limitations. For instance, the vacuum tubes used for the direct current circuit could not carry enough current or allow quick switches between "on" and "off" times.

Current and switching problems faded with the introduction of the transistor. Better accuracy and finishes resulted because the solid state device permitted the use of the proper current and switching for "on" and "off".

Today, EDM machines have enhanced servo systems, CNC-controls with fuzzy logic, automatic tool changers (Figure 2-3), and capabilities of simultaneous six-axes machining. EDM, along with wire EDM, has revolutionized machining.

去除破损的钻头和丝锥，这些早期的机器效率低下且难以使用。

后来，苏联科学家 Boris R. 和 Natalie I. Lazarenko 夫妇两人做了两个重大的改进。首先，他们开发了可控并可提供持续脉冲的 RC 电路；此外，他们开发了一套可以改善持续放电间隙，从而提供高效放电的伺服控制单元。

这两个改进使得电火花加工成为更可靠的加工手段，但是，这个过程仍然有它的局限性。比如，用在直流电路的真空管不能承载足够的电流，也不允许开关之间的快速切换。

随着晶体管的发明，电流和整流问题逐渐被解决了。由于可以使用更合适的电流和脉冲，电火花加工的精度更高了。

今天，电火花加工机器已经装备了伺服系统、具有模糊逻辑的 CNC 控制系统、工具自动转换装置（图 2-3）以及六轴联动能力的装置。电火花切削和电火花线切割已经掀起了加工的革命。

Figure 2-3　A CNC EDM with tool changer
自动换工具装置的数控电火花机器

2.2　How does EDM work　电火花加工机理

EDM uses spark erosion to remove metal. Its DC power supply generates electrical impulses between the workpiece and the electrode. A small gap between the electrode and the workpiece allows a flow of dielectric oil. When sufficient voltage is applied, the dielectric oil ionizes, and controlled sparks melt and vaporize the workpiece.

The pressurized dielectric oil cools the vaporized metal and removes the eroded material from the gap. A filter system cleans the suspended particles from the dielectric oil. The oil goes through a chiller to remove the generated heat from the spark erosion process. This chiller keeps the oil at a constant temperature which aids in machining accuracy. See Figure 2-4.

电火花加工利用火花放电腐蚀来去除金属材料。它的直流电源在工件和电极之间产生电流脉冲，在电极和工件的微小间隙之间允许流过工作液。当提供足够的电压时，工作液发生电离，受控的电火花熔化并汽化工件。

受压缩的工作液在工件和电极的缝隙中不断循环，不但冷却了被汽化的金属，并且冲洗走了被侵蚀的材料。在循环系统中有一个过滤系统用来过滤工作液中的悬浮颗粒。油液在过滤装置中不断循环往复，带走了电火花腐蚀加工过程产生的热量。该冷却装置使油液保持恒温，有助于提高加工精度，如图 2-4 所示。

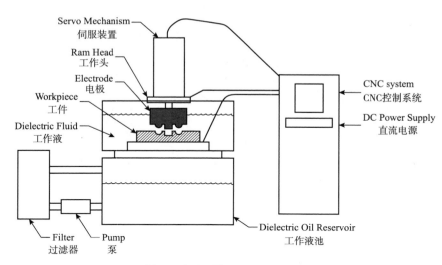

Figure 2-4　The EDM process
电火花加工过程

EDM is a spark erosion process. However, EDM produces the sparks along the surface of a formed electrode, as in Figure 2-5.

电火花加工是一个火花腐蚀的过程。然而，电火花加工是沿着成型电极的表面来产生火花的，如图 2-5 所示。

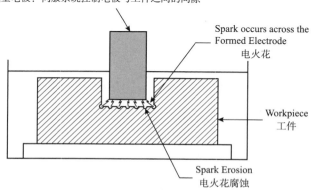

Figure 2-5　Spark erosion across the formed electrode
沿着成型电极产生电火花腐蚀

Former Electrode: A servo controls the gap between the electrode and the workpiece Spark occurs across the Formed Electrode.

A DC servo mechanism maintains the gap between the electrode and the workpiece from 0.001 to 0.002″(0.025 to 0.05mm). The servo system prevents the electrode from touching the workpiece. If the electrode were to touch the workpiece, it would create a short circuit and no cutting would occur.

2.2.1 The EDM Process

The DC power supply provides electric current to the electrode and the workpiece. (A positive or negative, charge is applied depending upon the desired cutting conditions.) The gap between the electrode and the workpiece is surrounded with dielectric oil. The oil acts as an insulator which allows sufficient current to develop. See Figure 2-6.

成型电极：伺服系统控制电极和工件之间的间隙，因此，电火花就发生在这两个成型电极之间。

电火花加工装置中的直流伺服机构能将电极和工件的间隙维持在 0.025~0.05mm。该伺服系统能有效地阻止电极接触工件。如果电极即将接触工件，伺服系统将很快发生短路来防止不能切削的情况发生。

2.2.1 电火花加工步骤

直流电源在电极和工件上产生电流（电极正接或反接，根据需要的切削情况来具体选择）。在电极和工件缝隙间充满的工作液作为一个绝缘体，以此来保证能有足够的电流产生，如图 2-6 所示。

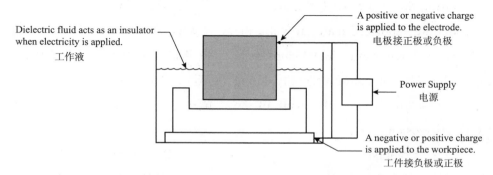

Figure 2-6　Power supply provides volts and amps
电源提供足够的电压和电流

Once sufficient electricity is applied to the electrode and the workpiece, the insulating properties of the dielectric oil break down, as in Figure 2-7. A plasma channel is quickly formed which reaches up to 8,000℃ to 12,000℃. The heat causes the fluid to ionize and allows sparks of sufficient intensity to melt and vaporize the material. This takes place during the controlled "on time" phase of the power supply.

一旦在电极和工件之间产生足够大的电流，绝缘性的油液将失去绝缘作用，如图 2-7 所示。此时，一个等离子通道将迅速形成，它的温度高达 14,500~22,000℉（8,000~12,000℃），这样的高温将引起油液的电离并产生足够强度的电火花来熔化并汽化材料。这个过程发生在电源的脉冲宽度范围内。

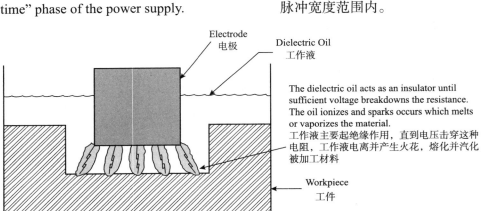

Figure 2-7　Sparks causes the material to melt and vaporize
电火花引起的材料熔化和汽化

During the "off times", the dielectric oil cools the vaporized material while the pressurized oil removes the EDM chips, as in Figure 2-8. The amount of electricity during the "on time" determines the depth of the workpiece erosion.

2.2.2　Polarity

Polarity refers to the direction of the current flow in relation to the electrode. The

在脉冲间隔范围内，工作液冷却被汽化的材料，同时高压油液将加工过程中产生的电蚀颗粒清除，如图 2-8 所示。在脉冲间隔中电流的大小决定了工件腐蚀的深度。

2.2.2　极性

极性是指在电极上产生的电流的流动方向，而电极的极性可

polarity can be either positive or negative.

Changing the polarity can have dramatic affects when Electrical discharge machining. Generally, electrodes with positive polarity wear better, while electrodes with negative polarity cut faster. However, some metals do not respond this way. Carbide, titanium, and copper should be cut with negative polarity.

2.2.3　No-Wear

An electrode that wears less than 1% is considered to be in the no-wear cycle. No-wear is achieved when the graphite electrode is in positive polarity and "on times" are long and "off times" are short. During the time of no-wear, the electrode will appear silvery showing that the workpiece is actually plating the electrode. During the no-wear cycle there is a danger that nodules will grow on the electrode, thereby changing its shape.

2.2.4　Fumes from the EDM Process

Fumes are emitted during the EDM process; therefore, a proper ventilation system should be installed. Boron carbide, titanium boride, and beryllium are three metals that give off toxic fumes when being machined; these metals need to be especially well-vented.

以是阳性也可以是阴性。

在电火花加工过程中，改变电极的极性可以极大地影响最后的加工结果。通常来说，阳极电极损耗得越快，阴极切削越快。但是，有些材料不采用这种反应，例如硬质合金、钛合金和铜被加工时应该接在阴极。

2.2.3　零磨损

当一个电极的磨损率低于1%时就可以认为它处于零磨损状态。当以石墨电极作为阳极且脉冲宽度比脉冲间隔长时就可以发生电极的零磨损。然而，零磨损电极的循环使用会使金属切削率有一个实质性的下降。当电极处于零磨损状态时，电极将呈现银白色，这表明工件正在电镀电极。在零磨损电极循环使用时，电极上会生长瘤状物，如此将改变电极的形状。

2.2.4　电火花加工过程中的烟气

在电火花加工过程中会产生烟气，因此，应当建立一套配套的通风系统。碳化硼、硼化钛、铍三种金属在电火花加工时会释放出有毒的烟气，所以需要谨慎地排放处理。

2.3 Dielectric Oil and Flushing for EDM 在电火花加工过程中的工作液及其冲洗作用

2.3.1 Dielectric Oil

EDM uses oil for its dielectric fluid. Dielectric oil performs three important functions for EDM, see Figure 2-8.

2.3.1 工作液

电火花加工使用油液作为它的绝缘流体。在电火花加工时，工作液发挥了三个重要作用，如图 2-8 所示。

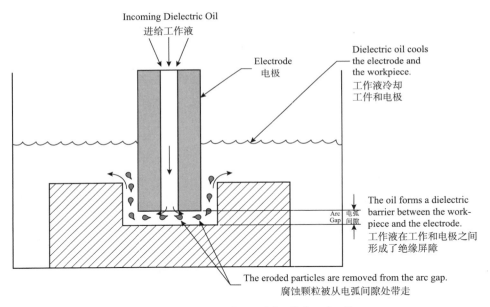

Figure 2-8 Functions of the dielectric oil
绝缘油的功能

(1) The fluid forms a dielectric barrier for the spark between the workpiece and the electrode.

(2) The fluid cools the eroded particles between the workpiece and the electrode.

（1）工作液在工件和电极之间形成了一道绝缘屏障。

（2）绝缘流体冷却了在工件和电极之间被腐蚀的颗粒。

(3) The pressurized fluid flushes out the eroded gap particles and removes the particles from the fluid by causing the fluid to pass through a filter system.

Many manufacturers produce many types of dielectric oil; the best way to determine the type of oil needed for a particular machine is to ask the machine manufacturer for its recommendations. It is important to get oil which is specifically produced for EDM.

2.3.2 Coolant System

EDM creates sparks in the gap with sufficient energy to melt the material. The resulting heat is transferred into the oil. Oil loses its efficiency when it reaches 100°F (38℃). Controlling this heat is essential to ensure accuracy and efficient cutting. Therefore, it is best to have a coolant system to maintain a proper temperature.

2.3.3 Flash Point

Oil will ignite at certain temperatures. The ignition temperature is called "flash point". This is especially important when doing heavy cutting, because the oil may get so hot that it reaches its flash point. Even though some oils have a flash point of 200°F (93℃) and higher, it is unsafe to use oil over

（3）高压油液冲洗掉了在工件和电极间隙之间被腐蚀的颗粒并且将油液中的电蚀颗粒通过过滤系统去除。

制造商们生产了很多类型的工作液。针对某台特定机器选择所需要类型的工作液最好的方法是询问机器制造厂商们的建议，因为针对电火花加工过程来说，工作液是非常重要的。

2.3.2 冷却液系统

电火花加工利用足够大的能量在工件与电极间产生火花来熔化金属。产生的热量最终被带入工作液中。当温度达到100°F（38℃）时，工作液将失去功效。合理地控制温度对保证加工精度和切削效率至关重要，因此，最好有一个冷却液系统来维持合适的温度。

2.3.3 引火点

工作液在某一特定温度将被点燃，这一燃烧温度被称为"引火点"。当进行大量的加工时，这一"引火点"的控制是非常重要的，因为工作液很可能因此而达到它的引火点。尽管某些油液的引火点达到200°F（93℃）或

165°F (74°C). Precautions need to be taken to prevent the oil from reaching its flash point. Some machines are equipped with a fire suppression system that is controlled by an infrared scanner.

2.3.4 Flushing

2.3.4.1 Proper Flushing

The most important factor in EDM is to have proper flushing. There is an old saying among EDMers, "There are three rules for successful EDMing: flushing, flushing, and flushing".

Flushing is important because eroded particles must be removed from the gap for efficient cutting. Flushing also brings fresh dielectric oil into the gap and cools the electrode and the workpiece. The deeper the cavity, the greater the difficulty for proper flushing.

Improper flushing causes erratic cutting. This in turn increases machining time. Under certain machining conditions, the eroded particles attach themselves to the workpiece. This prevents the electrode from cutting efficiently. It is then necessary to remove the

者更高，但是在加工过程中，只要油液温度超过165°F（74°C）都是不安全的。因此，必须采取适当的预防措施来阻止油液达到它的引火点。一些电火花加工设备就安装了配有红外线扫描仪的燃火抑制系统来控制此现象的发生。

2.3.4 油液的冲洗作用

2.3.4.1 合适的冲洗方法

在电火花加工过程中选择合理的冲洗方法是一个非常重要的因素。在电火花加工领域，有一句老话，"成功的电火花加工过程需要三个重要的角色：冲洗，冲洗，冲洗。"

能快速地将缝隙间的电蚀颗粒冲洗走对于高效切削来说作用巨大。冲洗作用也可以给缝隙间带来清洁的工作液以更好地冷却电极和工件。而且，加工腔体时，腔体越深，合适的冲洗方法就越为重要。

不适合的冲洗方法将引起不稳定的切削加工，而且会增加加工时间。在某种特定的加工情况下，电蚀颗粒将会附着在工件表面，对于切削效率的提高是一个非常不利的因素。因此，及时清

attached particles by cleaning the workpiece.

The danger of arcing in the gap also exists when the eroded particles have not been sufficiently removed. Arcing occurs when a portion of the cavity contains too many eroded particles and the electric current passes through the accumulated particles. This arcing causes an unwanted cavity or cavities which can destroy the workpiece. Arcing is most likely to occur during the finishing operation because of the small gap that is required for finishing. New power supplies have been developed to reduce this danger.

2.3.4.2 Volume, Not Pressure

Proper flushing depends on the volume of oil being flushed into the gap, rather than the flushing pressure. High flushing pressure can also cause excessive electrode wear by making the eroded particles bounce in the cavity. Generally, the ideal flushing pressure is between 3 to 5 psi. (0.2 to 0.33bars).

Efficient flushing requires a balance between volume and pressure. Roughing operations, where there is a much larger arc gap, requires high volume and low pressure for the proper oil flow. Finishing operations, where there is a small arc gap, requires higher pressure to ensure proper oil flow.

除和冲洗工件是非常必要的。

若缝隙间的电蚀颗粒没有被充分地冲走，将在缝隙间形成电弧，这是非常危险的。当腔体内存在过多的电蚀颗粒，电流就会通过这些积累在一起的颗粒，而产生电弧。电弧会形成有害的腔体从而严重破坏工件。电弧现象最有可能发生在精加工阶段，因为这时候电极和工件之间的间隙就要马上消失了，而新电源的使用能有效地减少这种危险。

2.3.4.2 体积而不是压强

合理的冲洗依靠的是冲入间隙间油液的体积而不是冲洗的压强。高压冲洗可引起电蚀颗粒在腔体内反弹，从而造成电极的过度磨损。通常来说，较为理想的冲洗压强为 3~5psi（0.2~0.33bar）。

高效切削要求冲洗液体积和压强之间的平衡。粗加工时，由于工件和电极的间隙较大，因此需要大体积低压强的油液来冲洗。精加工时，由于间隙较小，因此需要高压强来保证油液流通。

Often flushing is not a problem in a roughing cut because there is a sufficient gap for the coolant to flow. Flushing problems usually occur during finishing operations. The smaller gap makes it more difficult to achieve the proper oil flow to remove the eroded particles.

2.3.4.3 Types of Flushing

There are four types of flushing: pressure, suction, external, and pulse flushing. Each job needs to be evaluated to choose the best flushing method.

(1) Pressure Flushing

Pressure flushing, also called injection flushing, is the most common and preferred method for flushing. One great advantage of pressure flushing is that the operator can visually see the amount of oil that is being used for flushing. With pressure gauges, this method of flushing is simple to learn and use. Pressure flushing may be performed in two ways: through the electrode (Figure 2-9) or through the workpiece.

①Pressure Flushing through the Electrode

A problem with flushing through the electrode is when a stud or spike remains from the electrode flushing hole. If the stud gets too long, it can hinder proper flushing.

在粗加工时，冲洗往往不是主要的问题，因为有足够的间隙来确保冷却液流通。而精加工时，冲洗就成了一个麻烦的问题。过小的间隙很难保证足够的冷却液流通并冲走电蚀颗粒。

2.3.4.3 冲洗方式

常用的冲洗方式有四种：增压冲洗、抽吸冲洗、喷射冲洗和脉动冲洗。不同的加工需评估之后选择最好的冲洗方式。

（1）增压冲洗

增压冲洗也称作注射冲洗，它是最普通也是首选的冲洗方式。增压冲洗的优点是操作者能很直观地看见冲洗用的油量。因为有压力表指示，所以这种冲洗方法最容易学习和使用。增压冲洗可以通过两种途径来实现：油液通过电极（图2-9）和油液通过工件。

①通过电极的增压冲洗

使用通过电极的增压冲洗方式产生的一个问题是需要一个螺栓或螺钉来保持电极冲洗孔的通畅。如果螺栓太长，它可能会阻碍冲洗作用。

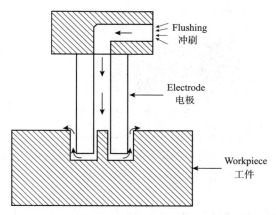

Figure 2-9　Pressure flushing through the electrode
　　　　　通过电极的增压冲洗

Occasionally, thin studs from soft metal weaken and touch the electrode and cause a short. These studs need to be removed either by hand, by a portable hand grinder, or with another electrode. Studs in hardened metal can be snapped off easily with needle nose pliers. If the machine has orbiting capabilities, the stud can be removed by orbiting.

Small holes are often drilled into large electrodes to aid in flushing, as in Figure 2-10. The placement of these holes is critical to ensure that flushing occurs over the entire cutting area. However, for certain applications, today's newer adaptive controls with fuzzy logic have reduced the need for flushing holes.

Flushing holes are drilled on an angle to prevent long studs from developing. See

软金属材料制成的细螺钉偶尔会触碰电极引起短路。这些螺钉可以通过人手、手动磨床或者另一个电极来移除。而用较硬的金属材料制成的螺钉可以通过尖嘴钳轻松剪断移除。如果机器能够旋转，也可以利用离心力将其甩出。

在冲洗过程中，往往会在大电极上钻很多小孔来辅助冲洗，如图 2-10 所示。这些孔的位置是极其严格的，以此来保证冲洗过程能在整个切削表面进行。但是，今天一些用模糊逻辑编程的软件的开发使用，已经能有效地减少冲洗孔的数量。

冲洗孔是以一定的角度钻在电极上以防止长螺钉从垂直孔下

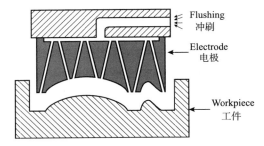

Figure 2-10 Multiple flush holes drilled through the electrode
钻在电极上的大量的冲洗孔

Figure 2-11. A disadvantage of angled holes is that they have a tendency to prevent proper flushing to occur by directing oil away from needed areas.

落。如图 2-11 所示。角度孔的一个缺点是它们改变了油流的方向，因此总是阻挡足够的油流来冲洗需要冲洗的区域。

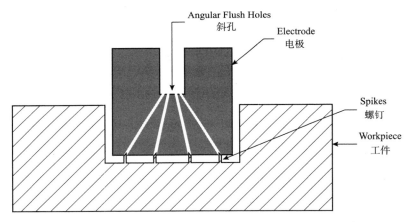

Figure 2-11 Tapering the Flush Holes Prevents Long Studs
逐渐减小冲洗孔来阻止长螺钉下落

When drilling stud holes, machinists often cover the roughing electrode with tape before drilling. They cover the holes with tape to prevent from drilling flushing holes in the same location for the finishing electrode. Holes are drilled in different locations so

在钻螺钉孔前，工人常常用胶布覆盖在粗糙的电极上，以此来防止冲洗孔钻在精细电极的位置上。孔钻在不同的位置以此可以确保精细电极上预先安置的螺钉能被取下。

that previous studs can be removed with the finishing electrode.

When electrodes are the same on both ends, the drilled flushing holes should be offset. By rotating the electrode 180 degrees, the previous studs are machined. See Figure 2-12.

当电极旋转 180°，之前的螺栓则都被电火花加工掉了，如图 2-12 所示。

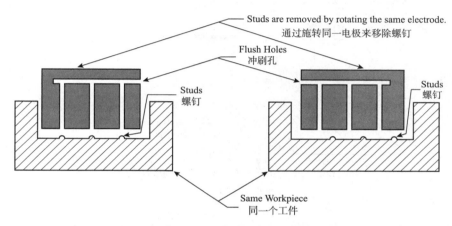

Figure 2-12　Rotating an electrode 180 degrees to remove studs
将电极旋转 180° 以去除螺栓

Stepped electrodes are sometimes used. If the flushing holes are uncovered in these stepped electrodes, the dielectric oil will escape and prevent flushing in the other holes. To insure the flushing process, a 0.120″ (3.05mm) hole is drilled, and a 1/8″ (3.18mm) diameter silicon tubing (commonly found in aquarium supply stores) is pressed into the electrode to seal the exposed hole. When that portion of the electrode is sealed, the tubing is pulled back so flushing can escape from the hole, as depicted in Figure 2-13.

我们偶尔会使用附加电极，如果这些附加电极上的冲洗孔没有被覆盖堵塞，电解质油将逸出阻碍其他孔的冲洗。为了保证冲洗过程顺利进行，需要钻一个直径 3.05mm 的冲洗孔，并将一根直径为 1/8″（3.18mm）的硅管（通常在水族馆用品店中可以找到）压入电极来密封暴露的孔。当电极的一部分被密封时，可以抽出硅管来使油液流通，如图 2-13 所示。

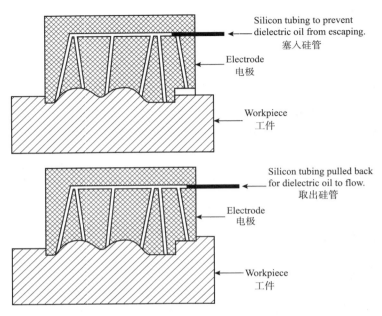

Figure 2-13　Flushing holes for stepped cavities
冲洗阶梯式孔

②Pressure Flushing through the Workpiece

Pressure flushing can also be done by forcing the dielectric fluid through a workpiece mounted over a flushing pot. See Figure 2-14. This method eliminates the need for holes in the electrode.

With pressure flushing, there is the danger of a secondary discharge. Since electricity takes the path of least resistance, secondary discharge machining can occur as the eroded particles pass between the walls of the electrode and the workpiece, as presented in Figure 2-15. This secondary discharge can

②穿过工件的增压冲洗

增压冲洗也可以采用工作液从安装好的冲洗池下穿过工件的方式进行，如图2-14所示。这种方式不需要在电极上打孔。

使用增压冲洗方式会发生二次放电现象。当电流通过的电阻过小，电极和工件之间的已加工表面由于电蚀产物的介入而发生二次放电从而产生二次加工现象，如图2-15所示。这种二次放电现象可能引起工件和电极相

Unit 2　Electrical discharge machining　电火花加工 | 029

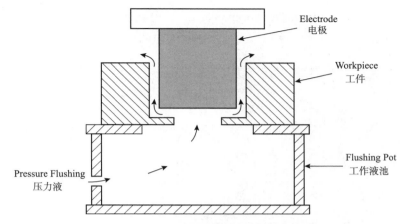

Figure 2-14　Pressure flushing through the workpiece
穿过工件的增压冲洗

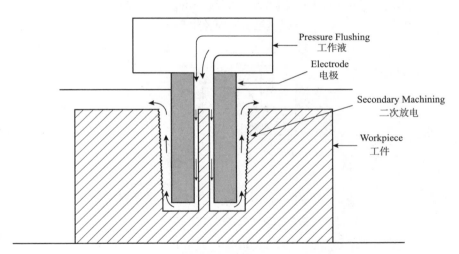

Figure 2-15　Pressure flushing may cause secondary machining
增压冲洗可能引起二次放电

cause side wall tapering. Suction flushing can prevent side wall tapering.

(2) Suction Flushing

Suction or vacuum flushing can be used to remove eroded gap particles. Suction flushing can be done through the electrode as

对的表面逐渐减小，而抽吸冲洗则可以阻止该现象发生。

（2）抽吸冲洗

抽吸或者真空冲洗也被用来清除工件和电极之间的电蚀颗粒。抽吸冲洗可以利用油液穿过

in Figure 2-16, or through the workpiece, as in Figure 2-17.

电极或者穿过工件的方式来完成，如图 2-16 和图 2-17 所示。

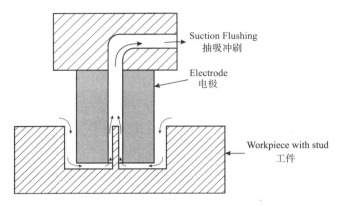

Figure 2-16　Suction flushing through the electrode
穿过电极的抽吸冲洗

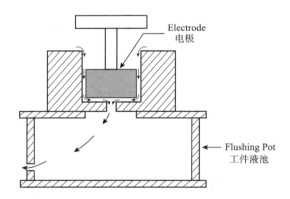

Figure 2-17　Suction flushing through the workpiece
穿过工件的抽吸冲洗

Suction flushing minimizes secondary discharge and wall tapering. Suction flushing sucks oil from the worktank, not from the clean filtered oil as in pressure flushing. For suction cutting, efficient cutting is best accomplished when the worktank oil is clean.

抽吸冲洗最大限度地减少了二次放电和二次加工现象。抽吸冲洗从工作池中吸取油液而不是像增压冲洗一样从过滤装置中汲取。使用抽吸冲洗，如果工作池中的油液纯净，高效切割可以很好地实现。

A disadvantage of suction flushing is that there is no visible oil stream as with pressure flushing. Also, gauge readings are not always reliable regarding the actual flushing pressure in the gap.

A danger of suction flushing is that gases may not be sufficiently removed, this can cause the electrode to explode. In addition, the created vacuum can be so great that the electrode can be pulled from its mount, or the workpiece pulled from the magnetic chuck.

(3) Combined Pressure and Suction Flushing

Pressure and suction flushing can be combined. They are often used for molds with complex shapes. This combination method allows gases and eroded particles in convex shapes to leave the area and permit circulation for proper machining.

(4) Jet Flushing

Jet or side flushing is done by tubes or flushing nozzles which direct the dielectric fluid into the gap, as shown in Figure 2-18.

Although jet flushing is a convenient method of flushing, and sometimes the only choice, it is also the most ineffective way to remove eroded gap particles. The danger of not removing the gap particles is that DC arcing can occur. When placing the nozzles

抽吸冲洗的一个缺点是没有像增压冲洗一样明显可见的油流。同样，对于间隙中的实际压强，压力表的读数并不总是准确的。

抽吸冲洗可能发生的危险是加工产生的气体不能充分地被排出，这可能造成电极爆炸。除此之外，安装的真空装置过大可能会导致电极从底座上被拔出，或者工件从磁性卡盘上被拔出。

（3）增压和抽吸冲洗的联合使用

增压冲洗和抽吸冲洗可以联合使用，常常用在加工复杂形状模具时。这种联合方式可以使加工产生的气体和电蚀颗粒排出，有利于加工过程的循环进行。

（4）喷射冲洗

喷射冲洗是利用管子或冲洗喷嘴直接将工作液射入工件与电极间隙的方式，如图2-18所示。

尽管喷射冲洗是一种方便的冲洗方式，但是有时这种唯一的选择却是去除缝隙间电蚀颗粒最无效的方式，不消除间隙粒子会产生危险的直流电弧。当安装喷射冲洗的喷嘴时，油液必须

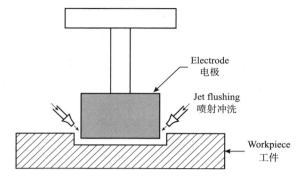

Figure 2-18　Jet flushing using multiple flushing nozzles
喷射冲洗利用大量的冲洗喷嘴

for jet flushing, the fluid must be directed so it can remove the eroded particles from the gap.

One advantage of jet flushing, however, is that it leaves no stud. For shallow cuts, this is an effective method. But as depth increases, external flushing decreases in its effectiveness. Pulse flushing is usually used along with jet flushing.

定向，以此来有效地去除电蚀颗粒。

然而，喷射冲洗的一个优点是它不会留下螺钉。对于较浅的电火花加工，这是一种有效的冲洗方式。但是一旦深度增加，喷射冲洗的效果就会减弱。脉冲冲洗通常与喷射冲洗一起使用。

2.4　EDM Applications　电火花加工的应用

EDM is a relatively mature non-traditional machining method, and some EDM applications are listed as follows: Small holes processing (Figure 2-19). Deep holes processing (Figure 2-20). Square holes processing (Figure 2-21)

电火花加工是特种加工中一种相对成熟的加工方式，以下是一些常用的电火花加工应用：小孔加工（图2-19）、深孔加工（图2-20）、方孔加工（图2-21）。

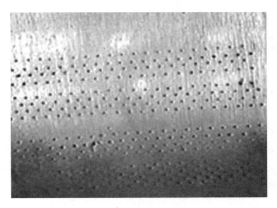

Figure 2-19　Cutting small holes for the aero-engine
飞机引擎上的小孔加工

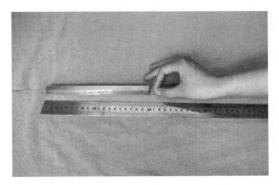

Figure 2-20　Cutting a deep hole
深孔加工

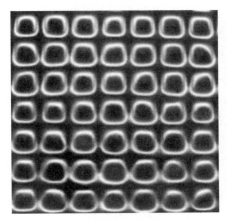

Figure 2-21　Cutting a square hole
方孔加工

Questions 习题

1. List the various names of EDM.
2. List and explain the two significant improvements in spark erosion from the two Russian scientists.
3. How does the EDM work?
4. What function does the dielectric oil have when electricity is supplied?
5. What happens when sufficient electricity is supplied between the electrode and the workpiece?
6. What happens during the off time of the electrical cycle?
7. How many types of flushing between the electrode and the workpiece?
8. What kinds of materials can be machined with EDM?
9. List three applications for EDM.

1. 列出电火花加工的各种名称。
2. 列出并解释两位俄国科学家在电火花腐蚀上进行的两个重要改进。
3. 简述电火花加工的机理。
4. 进行电火花加工时工作液的作用是什么？
5. 在电极和工件之间通以足够大的电流时会发生什么现象？
6. 脉冲循环过程中在脉冲间隔内会发生什么现象？
7. 电极和工件之间的冲洗方式有多少种？
8. 哪些材料可以用电火花加工来加工？
9. 列出三种电火花加工的应用。

Wire Electrical Discharge Machining 电火花线切割加工 3

3.1 Fundamentals of Wire Electrical Discharge Machining 电火花线切割加工基础

Wire Electrical Discharge Machining (WEDM), as shown in Figure 3-1, is one of the greatest innovations affecting the tooling and machining industry. This process has brought to industry dramatic improvements in accuracy, quality, productivity, and earnings.

电火花线切割加工（WEDM）如图 3-1 所示，是极大影响了加工工业的一项伟大创新，此方法已经在精确度、加工质量、生产率和利润方面给加工工业带来了巨大的进步。

Figure 3-1　Wire electrical discharge machining
电火花线切割

Before Wire EDM, costly processes were often used to produce finished parts. Now with the aid of a computer and Wire EDM machines, extremely complicated shapes can be cut automatically, precisely, and economically, even in materials as hard as carbide.

In 1969, the Swiss firm Agie produced the world's first Wire EDM machine. Typically, these first machines in the early 70s' were extremely slow, cutting about 2 square inches an hour ($21mm^2$/min). Their speeds went up in the early 80s' to 6 square inches an hour ($64mm^2$/min). Today, machines are equipped with automatic wire threading and can cut nearly 30 square inches an hour ($322mm^2$/min).

Some machines cut to accuracies of up to ±0.0025mm, producing surface finishes to Ra 0.037μm and lower. One Wire EDM machine on the market uses a special cutting fluid to produce a mirror-like finish.

Wire EDM competes seriously with such conventional machining as milling, broaching, grinding, stamping, and fine-blanking. Conventional wisdom suggests that

在电火花线切割出现之前，成型工件的加工往往使用一些成本较高的加工方法。现在，在电脑和电火花线切割机床的辅助下，极其复杂的形状也可以被自动、准确、经济地切割出来，甚至连金刚石这种极其坚硬的材料也能被切割成型。

1969年，Swiss firm Agie 发明了世界上第一台电火花线切割机床。但在19世纪70年代，第一代电火花线切割机床的加工速度是十分慢的，加工速度大约为$2in^2$/h（约$21mm^2$/min）。80年代时，该速度上升到$6in^2$/h（约$64 mm^2$/min）。如今，电火花线切割机床装备有自动卷丝装置，并能以约$30in^2$/h（约$322mm^2$/min）的速度进给加工。

某些机器加工的产品可以达±0.0025mm 的精度，表面粗糙度也可以达到Ra 0.037 μm 或者更低的值。市面上的电火花线切割机床使用一种特殊的切削液，能够加工出类似镜面的表面。

电火花线切割加工和磨削加工、钻削、磨粒加工、板料冲压和精密冲裁等传统加工方式竞争十分激烈。传统观点认为WEDM

Wire EDM is only competitive when dealing with expensive and difficult-to-machine parts. But this is not the case. Wire EDM is often used with simple shapes and easily machined materials.As more design engineers discover many advantages of Wire EDM, they are incorporating new designs into their drawings. It therefore becomes important for contract shops to understand Wire EDM so they can properly quote on these new designs requiring EDM.

Increasingly, today's drawings are calling for tighter tolerances, shapes that only can be efficiently machined with Wire EDM, and alloys difficult to machine, as illustrated in Figure 3-2. With Wire EDM these exotic alloys can be machined just as easily as mild steel. When Wire EDM manufacturers select the optimum steel to demonstrate the

只有在处理昂贵材料和难加工的工件时才有竞争力，但是事实并非如此，WEDM 也经常用于一些简单形状的零件和容易加工的材料的加工。随着越来越多的设计者发现 WEDM 的诸多优势，他们也将许多新的设计加入设计图纸中去。因此对合约方来说，懂得 WEDM 的加工方式便十分重要，这样才能合适地采用一些需要用 WEDM 的新设计方案。

如今，越来越多的设计要求高精度等级的公差，有些形状和某些难加工的合金只有用 WEDM 才能加工。如图 3-2 所示，使用 WEDM 的方式加工这些特殊的合金便可以如同普通的钢一般容易。当 WEDM 厂商选择适宜的钢件来展示他们机器的能力时，

Figure 3-2　Some difficult-to-machine shapes done with wire EDM
　　　　　一些用 WEDM 加工的难加工工件

capability of their machines, their choice is not mild steel, but hardened D2, a high-chrome, high-carbon tool steel.

他们往往不选择普通的钢，而是选择硬质合金块作为加工材料。

3.2 How does Wire EDM work

WEDM 的工作过程

Wire EDM uses a traveling wire electrode that passes through the workpiece. The wire is monitored precisely by a computer-numerically controlled (CNC) system. See Figure 3-3.

WEDM 使用移动着穿过工件的线电极，电极丝的进给由 CNC 系统精密地控制，如图 3-3 所示。

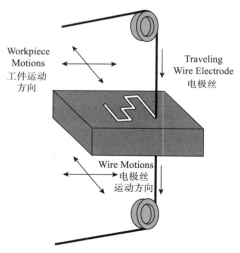

Figure 3-3　Wire EDM
电火花线切割加工

Like any other machining tool, Wire EDM removes metal; but Wire EDM removes metal from the workpiece with electricity by means of spark erosion.

像其他加工方式一样，电火花线切割加工也是移除材料，但是 WEDM 是利用电能依靠电火花放电腐蚀材料的方式移除材料的。

Rapid DC electrical pulses are generated between the wire electrode and the workpiece. Between the wire and the workpiece there is a shield of deionized water, called the dielectric. Pure water is an insulator, but tap water usually contains minerals that causes the water to be too conductive for Wire EDM. To control the water conductivity, the water goes through a resin tank to remove much of its conductive elements-this is called deionized water. As the machine cuts, the conductivity of the water tends to rise, and a pump automatically forces the water through the resin tank when the conductivity of the water is too high.

When sufficient voltage is applied, the fluid ionizes. Then a controlled spark precisely erodes a small section of the workpiece, causing it to melt and vaporize. These electrical pulses are repeated thousands of times per second. The pressurized cooling fluid, the dielectric, cools the vaporized metal and forces the re-solidified eroded particles from the gap.

The dielectric fluid goes through a filter which removes the suspended solids. To maintain machine accuracy, the dielectric fluid flows through a chiller to keep the liquid at a constant temperature.

在电极丝和工件之间会形成直流电脉冲。电极丝和工件之间存在着去离子水屏障，称为电介质。纯水是绝缘体，但是由于自来水中含有矿物质能导电，因而可以作为 WEDM 的电介质。为了控制导电性，通常将作为电介质的水通过合成树脂塔，从而移除大多数导电的离子，得到去离子水。当加工切割进行时，介质的导电性会增加，为了避免其导电性过高，通常在加工过程中用泵将电介质抽出，注入树脂塔中过滤去除离子，形成电介质在工件和合成树脂塔中的循环。

当外加电压足够高的时候，电介质将电解，然后可控的火花放电精确地移除工件的一小部分，并将其熔化汽化。这种电脉冲每秒将重复数千次，同时高压冷却电介质将汽化的材料冷却并使重新固化的腐蚀颗粒移出加工间隙。

电介质通过过滤器滤除被去除的工件颗粒。为了保持加工精度，电介质将通过冷却器来保持恒温。

A DC or AC servo system maintains a gap from 0.001 to 0.002 between the electrode and the workpiece. The servo mechanism prevents the wire electrode from shorting out against the workpiece and advances the machine as it cuts the desired shape.

The wire electrode is usually a spool of brass, or brass and zinc wire from 0.05mm to 0.33mm thick. Sometimes molybdenum wire is used. Some machines can even cut with 0.025mm wire. To cut with such thin wires, tungsten is used. New wire is constantly used which accounts for the extreme accuracy of Wire EDM.

3.2.1 The Step by Step Wire EDM Process

(1) Power supply generates volts: Deionized water surrounds the wire electrode as the power supply generates volts to produce the spark.

(2) During on time, the controlled spark erodes materials: The generated spark precisely melts and vaporizes the material.

(3) Off time allows the fluid to remove eroded particles: During the off cycle, the pressurized dielectric fluid immediately cools the material and flushes the eroded particles.

(4) Filter system removes chips while

直流或交流伺服系统控制工具和工件之间的加工间隙在 0.001~0.002 mm。伺服系统可以保证加工过程中电极丝和工件之间的间隙，防止短路现象发生，并加快加工过程。

电极丝通常为绕在储丝筒上面的黄铜丝，或者是黄铜－锌合金，直径为 0.05~0.33 mm 不等。有时候电极丝也使用钼丝。有些机床的电极丝直径甚至为 0.025 mm，此时电极丝使用钨丝。当对 WEDM 加工工件的精度要求高时，电极丝通常不重复使用。

3.2.1 WEDM 的加工步骤

（1）脉冲电源产生电压：当脉冲电压产生时，去离子水电离以产生电火花。

（2）脉冲宽度内电火花腐蚀工件：电火花将精确地熔化汽化工件材料。

（3）脉冲间隔内，电介质清除腐蚀颗粒：在脉冲间隔内，高压去离子水冷却工件材料，并带走腐蚀颗粒。

（4）循环过程中，过滤器滤

the cycle is repeated: New wire is constantly fed, while the eroded particles are removed and separated by a filter system.

3.2.2 Three Types of Wire EDM

(1) Two Axis

Two axis Wire EDM permits only right angle cuts. See Figure 3-4.

除腐蚀颗粒：在腐蚀颗粒被过滤器滤除的同时，电极丝也在循环转动。

3.2.2 三种电火花线切割加工机床

（1）两轴式

两轴式 WEDM 机床只能进行直角切削，如图 3-4 所示。

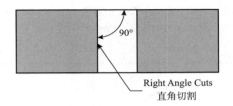

Figure 3-4　Two axis-For right angle cuts only.
两轴式电火花线切割直角加工

(2) Simultaneous Four Axis

Simultaneous four axis machines can produce a taper that is the same shape both on the top and on the bottom, as pictured in Figure 3-5. They are also capable of going from a straight cut to a taper. This type of machine is particularly useful for making stamping dies. A straight land with a taper can also be produced with a secondary skim cut.

(3) Independent Four Axis

Independent four axis machines can cut a top profile different from the bottom profile. See Figure 3-6. This is particularly useful for extrusion molds and flow valves.

（2）四轴联动式

四轴联动式机床可以切削出一个上、下同形的锥面，如图 3-5 所示；也可以由直线切削过渡到锥面切削。此类机床通常用于加工冲压模具。当要求切削柱面和锥面连接在一起的形状时，可以进行二度精切。

（3）四轴独动型

四轴独动型机床可以切削出顶部和底部不同形状的工件，如图 3-6 所示，通常用于冲压模型和流量阀加工。

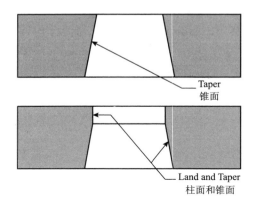

Figure 3-5　Simultaneous four axis-For taper and straight cuts.
四轴联动型机床切削锥面和柱面

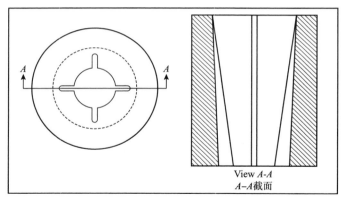

Figure 3-6　Independent four axis
Different shapes can be produced on top and bottom of a workpiece
四轴独动型
工件上、下部可以有很大形状区别

3.2.3　Materials That Can be Wire EDMed

Wire EDM can cut any electrically-conductive hard or soft material. Some of the materials that can be Wire EDMed are listed below.

3.2.3　可以用 WEDM 切削的材料

WEDM 可以切削任何导电的材料，无论其硬度如何。一些常用的可以用 WEDM 切削的材料如表 3-1 所示。

Table 3-1 Materials that can be Wire EDMed 可以用 WEDM 加工的材料

Aluminum 铝	Tool Steels: A2,D2,S7 工具钢 : A2,D2,S7	PCD Diamond 导电金刚石
Copper 铜	Stainless Steels 不锈钢	Nickel 镍
Titanium 钛	Carbide 碳化物	Stellite 合金

3.3 Proper Procedures WEDM 的加工步骤

To gain the greatest benefits from Wire EDM, specific procedures should be used to maximize EDM's potential for reducing machining costs. In planning work, the Wire EDM machine can be visualized as a super precision band saw which can cut any hard or soft electrical conductive material.

3.3.1 Starting Methods for Edges and Holes

If the outside edges are important, then a finished edge should be indicated when setting up the part to be Wire EDMed.

（1）Pick Up Two Edges as in Figure 3-7.

（2）Pick Up a Hole as in Figure 3-8.

3.3.2 Cutting Speed

Speed is rated by the square inches of material that are cut in one hour. Manufacturers rate their equipment under

若要从 WEDM 加工中获得最大的利益，应采用特殊的工艺发挥 WEDM 最大的潜能，以减少加工费用。在加工计划中，WEDM 机床应视为一种高精度的条形锯，用 WEDM 的方法可以切削任何硬度的导电材料。

3.3.1 加工轮廓和孔时的启动方式

如果外轮廓很重要，那么在设置采用 WEDM 加工工件时，外轮廓的终止处应被指定。

（1）选定两个轮廓边界，如图 3-7 所示。

（2）加工出一个孔，如图 3-8 所示。

3.3.2 切削速度

WEDM 的切削速度以每小时切削材料面积的平方英寸数目来定义。设定加工速度时，通常以

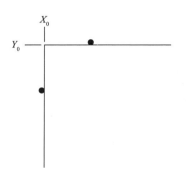

Figure 3-7　Pick up two edges
　　　　　选定两个轮廓边界

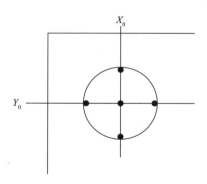

Figure 3-8　Pick up a hole
　　　　　加工出一个孔

ideal conditions, usually $2\tfrac{1}{4}$ inch (57mm) thick D2 hardened tool steel under perfect flushing conditions. However, differences in thicknesses, materials, and required accuracies can greatly alter the speeds of EDM machines.

Cutting speed varies according to the conductivity and the melting properties of materials. For example, aluminum, a good conductor with a low melting temperature, cuts much faster than steel.

On the other hand, carbide, a non-conductor, cuts much slower than steel. It is the binder, usually cobalt, that is melted away. When the cobalt is eroded, it causes the carbides to fall out. Various carbides machine at different speeds because of carbide grain size and the binder amount and type.

$2\tfrac{1}{4}''$（70mm）厚的 D2 硬质工具钢在良好的电解液供给下的速率来定义。然而工件厚度、材质、要求加工精度都可能对 WEDM 的加工速率产生很大影响。

切削速度也因材料的导电性和材料的熔点不同而不同。例如：铝的导电性强于钢，而熔点低于钢，其加工速度便比钢的加工速度快。

另一方面，碳化物是绝缘体，其切削速度比钢慢。通常是其中的黏合剂钴被熔化。当钴被腐蚀时，将导致碳化物上面开始掉落细小的颗粒。不同碳化物加工速度不同的原因在于碳晶粒的尺寸不同以及黏合物含量及其种类不同。

3.3.3 AC Non-Electrolysis Power Supplies

Instead of cutting with DC (direct current), some machines cut with AC (alternating current). Cutting with AC allows more heat to be absorbed by the wire instead of the workpiece.

Since AC constantly reverses the polarity of the electrical current, it reduces the heat-affected zone and eliminates electrolysis. Electrolysis is the stray electrical current that occurs when cutting with Wire EDM. For most purposes, electrolysis does not have any significant affect on the material. However, the elimination of electrolysis is particularly beneficial when cutting precision carbide dies in that it reduces cobalt depletion.

When titanium is cut with a DC power supply, there is a blue color along where the material was cut. This blue line is not caused by heat, but by electrolysis. As some suspect, this effect is not generally detrimental to the material. However, AC power supply eliminates this line.

Like AC power supply, the AE (anti-electrolysis) or EF (electrolysis-free) power supplies improve the surface finish of parts by reducing rust and oxidizing effects of

3.3.3 交流无损耗电源

有些机床使用交流电源（电流方向交替变化的脉冲电源）而不是直流电源（电流方向不变的电源）。使用交流电源加工时允许更多热量被工具电极吸收而不是被工件吸收。

由于交流电源连续改变电流方向，这便减少了热影响区范围并降低了加工损耗。WEDM 的加工损耗是零散的电脉冲引起的。绝大多数情况下，电腐蚀损耗对加工没有重大影响。然而，电腐蚀量的降低对碳化物模具的精确加工是有很大好处的，这是因为 WEDM 可以减少钴的消耗量。

当电源为直流电源，使用 WEDM 的方式切削钴的时候，材料被切削的地方会有蓝色的电火花带产生。这条蓝色的电火花线不是由热产生的，而是由电蚀产生。如同专家推测，这种效果通常对材料无伤害。然而，使用交流电源却可以避免这条蓝线的产生。

如同交流电源一样，反电蚀电源或无电蚀电源提高了工件加工面的表面质量，这种质量的提升是通过减少 WEDM 的加工锈蚀

Wire EDM. Also, less cobalt binder depletion occurs when cutting carbide, and it eliminates the production of blue lines when cutting titanium. AC and non-electrolysis power supplies definitely have advantages.

3.3.4 Heat–Treated Steels

Wire EDM will machine hard or soft steel; however, steel in the hardened condition cuts slightly faster. Materials requiring hardening are commonly heat treated before being cut with wire. By heat treating steel beforehand, it eliminates the distortions that can be created from heat-treating.

3.3.5 Cutting Large Sections

Steels from mills have inherent stresses. Even hardened steel that has been tempered often has stresses remaining. For cutting small sections, the effect is negligible. However, for large sections when there is a danger of metal movement, it is advisable to remove some of the metal. By removing metal, it reduces the possibility of metal movement. See Figure 3-9.

和氧化效应而实现的。同样，当切削碳化物时，钴黏合剂的消耗量也大大减少，切削钴时的蓝色光带也随之消除。交流电源和无损耗电源确实有着加工上的优势。

3.3.4 热处理钢

WEDM 可以用来切削任何硬度的钢材，然而，硬质钢材的切削速度要稍快一些。有硬度要求的材料通常在 WEDM 加工之前进行热处理。通过对钢材预先热处理，可以消除因加工后热处理而产生的变形。

3.3.5 大尺寸工件的切削

磨削之后的钢材会有内应力存在，即使是已经回火的高硬度钢材通常也会有残余应力。对于小件的切削，这种影响是可以忽略不计的。而当切削大尺寸工件时，当有金属变形的风险时，更有效的做法是切削掉一些金属。通过对金属的切削，可以减少金属变形的风险，如图 3-9 所示。

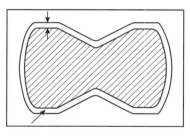

Figure 3-9 Removing material to reduce stress on large parts
切削部分金属以减少大尺寸工件的应力

3.4 Advantages of Wire EDM for Die Making
WEDM 在模具制造方面的优势

A great problem is that many designers and engineers are unaware of the great potential of Wire EDM to machine parts. As more and more engineers, tool designers, and machinists discover the capabilities of Wire EDM, they discover that Wire EDM is a very cost effective and accurate machining process. One of the major tasks is to educate industry in the capabilities and advantages of Wire EDM.

To produce these precision dies, it required highly skilled tool and die makers. Then came Wire EDM. Now making a computer program of the shape, the production of a much better and more accurate tool was possible. Tool and die makers are still needed to assemble tooling, but Wire EDM has eliminated the need for those skilled die makers to make the many elaborate punch and die sections. Today, Wire EDM performs that costly and laborious job. As a result, it has greatly reduced tooling costs, and at the same time produced a superior quality die.

The advantages of Wire EDM for die

如今，加工工业存在一个很大的问题，那就是许多设计者和工程师都没有意识到 WEDM 在加工工件方面的巨大潜力。当越来越多的工程师、设计师、机械技工发现 WEDM 的潜质，他们发现 WEDM 是一种十分经济且精确的加工方式。我们的主要任务就是增强工厂对于 WEDM 加工的巨大潜力和优势的意识。

若要生产高精度的模具，必须要有熟练的工具模具生产工人，这便体现出了 WEDM 的优势。现在，使用电脑对工件的形状进行编程，可加工出更好、更高精度的工具。工具和模具的生产厂商仍然要组装工具，但是 WEDM 便减少了对于工人熟练掌握制造精密组件和模具组件技能的要求。如今，WEDM 执行了这项花费大而工时长的工作。因此，它极大地减少了工具加工的花费，同时还可以生产出更多高质量的模具。

WEDM生产模具的优点如下：

making are listed as follows:

3.4.1 One-Piece Die Sections

Previously complicated dies were sectionalized-this allowed the die sections to move. See Figure 3-10(a). Now with Wire EDM, the die can be made from a solid block of tool steel producing a much more rigid die, as in Figure 3-10(b). In addition, sectionalized dies require much more mounting time than one-piece die sections.

3.4.1 整体模具生产

以前的复杂模具生产通常是部件分开生产的，这就会产生部件之间的相互错动，如图3-10（a）所示。现在随着WEDM的使用，这种模具可以由一整块工具钢切削而成，这样切削出的模具具有更高的刚度，如图3-10（b）所示。另外，多组件的模具与整体模具相比，前者要花费更多的安装时间。

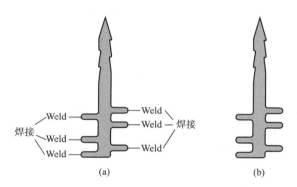

Figure 3-10　Sectionalized die sections and solid die section
多组件组合式模具和单组件整体式模具

3.4.2 Exact Spare Parts

To keep up production, spare sections can be on hand in case of wear or breakage. Since computer programs can be stored, spare sections can be precisely duplicated without having the previous part.

3.4.2 精确制作备用零件

为了使生产率保持在高水平上，必须时刻备有备用零部件以更换磨损严重及破损零件。由于计算机程序可以被储存起来，所以备用零部件可以在没有模板的情况下精确地制出。

3.4.3 Dowel Holes EDMed

When die sections or punches need to be changed due to wear or design change, dowel holes can also be EDMed. This produces exactly duplicated replacement die or punch sections.

3.4.4 Better Tool Steels

With Wire EDM, dies and punches can be made with tougher tool steels, even tungsten carbide. These tougher tool steels produce much longer tool life.

3.4.5 Accuracy

Many Wire EDM machines move in increments of at least 40 millions of an inch, therefore they can maintain accurate forms and clearances.

3.4.6 Die Repairs

Broken dies can be saved by replacing the damaged section With a Wire EDMed insert, or the damaged area can be hard welded and then Wire EDMed. See Figure 3-11 and Figure 3-12.

3.4.7 Fine Textured Finish

The fine textured surface produced from Wire EDM produces longer tool life because of improved surface retention of lubricant.

3.4.3 电火花加工定位销

当模板部件或者孔因过度磨损或设计改动要被换掉的时候，小孔也可以使用电火花加工的方式加工，这种生产精确地复制了替换品模具或孔类部件。

3.4.4 使用更好的工具钢作为材料

利用 WEDM，模具和孔可以用更坚硬的工具钢来制造，甚至是碳化钨，这种坚硬的工具钢具有更长的寿命。

3.4.5 精度

许多 WEDM 机床可以以每英寸至少 4×10^6 次的进给量加工，因此被加工工件可以保持精确的形状和良好的粗糙度。

3.4.6 模具更换

毁坏的模具可以通过使用 WEDM 方式加工出的部件更换损坏部分的方式来修复，或者是将损坏的部件焊接之后使用 WEDM 整边，如图 3-11 和图 3-12 所示。

3.4.7 良好的表面质量

由于 WEDM 提高了工件的表面保持能力和润滑性，因此加工出的工件表面质量好，并且有较长的寿命。

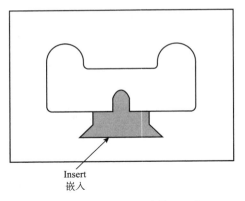

Figure 3-11 Damaged die section repaired with an insert
使用新部件更换损坏部件

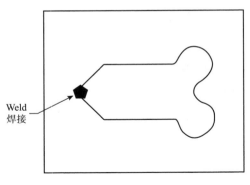

Figure 3-12 Damaged die section repaired with welding and EDMing
用焊接和 WEDM 的方式修复受损部件

3.4.8 Eliminates Distortion

Punch and dies can be Wire EDMed after heat-treatment. This eliminates the distortions that are created in heat-treating.

3.4.9 Inserts for High Wear Areas

If certain areas in the die have a larger wear ratio, inserts can be designed for these wear areas. Then instead of sharpening the entire die, inserts can be installed even with the die in the press.

3.4.10 Smaller Dies

Wire EDM allows the building of smaller progressive dies, thereby reducing costs.

3.4.11 Longer Lasting

A die lasts only as long as its weakest

3.4.8 减少失真现象

孔和模具可以在预先热处理的基础上以 WEDM 的方式加工，这样便消除了在加工后热处理有可能造成的变形。

3.4.9 高磨损区域的嵌块

如果工具上面的某处区域有较高的磨损速率，可以将该区域设计成嵌块形式。这样，当工具磨损时不必更换整个工具，而是仅仅更换磨损的嵌块。即使是工具处于受载情况下也可以更换。

3.4.10 更小的工具

WEDM 允许使用更小的工具，这样便减少了成本。

3.4.11 长寿命

一把刀具能工作的寿命与其

link. Dies last longer because Wire EDM produces exact die clearance which allows the dies to last longer between sharpening, and to be sharpened much farther into the die sections.

最薄弱环节的寿命一致。WEDM 的工具寿命更长，因为 WEDM 的生产过程中刀具与工件之间存在间隙，从而允许刀具工作更长时间而不会产生过度磨损现象，而且还允许刀具有更大的磨损量。

3.4.12 Punches and Dies From One Piece of Tool Steel

A punch and die can be produced from one piece of tool steel. See Figure 3-13.

3.4.12 在同一块工具钢上打孔并冲模

可以在同一块工具钢上打孔并冲模，如图 3-13 所示。

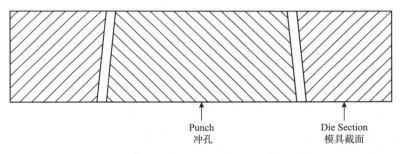

Figure 3-13　Punch and die made from one piece of tool steel
在同一块工具钢上打孔并冲模

3.4.13 Cutting Stripper and Die Section Together

Often the stripper may be mounted on the bottom of the die section and cut simultaneously with the die section as shown in Figure 3-14.This significantly reduces the cost when strippers are required.

3.4.13 同时切削脱料背板和模具部件

通常脱料背板会被放置于模具部件的下部并与模具一起切削，如图 3-14 所示。这种加工方式在需要脱料背板的时候能节省很多成本。

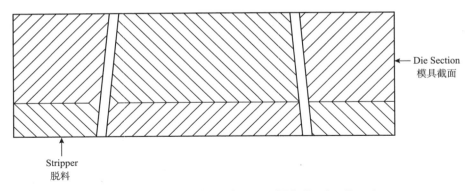

Figure 3-14　Cutting the Stripper and Die Section Together
同时切削脱料背板和模具部件

3.5　Wire EDM Applications　WEDM 的应用

Understanding the capabilities of Wire EDM permits many unique applications. Following are examples of work done by Wire EDM.

3.5.1　Cutting Tall Parts

Four angular cutouts were made on a large shaft, see Figure 3-15.

3.5.2　Modified Machines

We also did a slight modification on another machine. See Figure 3-16.

3.5.3　Large Heavy Gears Cutting

Cutting two keyways into a 54 inch (1,372mm) diameter gear weighing 3,000 pounds. See Figure 3-17.

了解了 WEDM 的众多优越性，我们很自然地便可以想到 WEDM 可以用在很多特殊的场合。以下便是用 WEDM 加工的具体实例。

3.5.1　切削长工件

图 3-15 为四角切削一根大的轴类部件。

3.5.2　修理机床

我们可以对其他机床做小的修整，如图 3-16 所示。

3.5.3　大、重齿轮的加工

如图 3-17 所示，在直径为 54in（1,372mm）、重量为 3,000lb 的齿轮上切削两个键槽。

Figure 3-15　Four angular cutouts were made on a large shaft
四角切削一根大的轴类部件

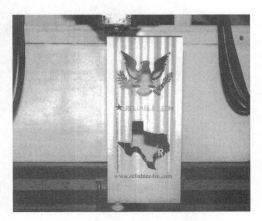

Figure 3-16　Machines Modified
机器修整

Figure 3-17　Cutting two keyways into a 54 inch (1,372mm) diameter gear weighing 3,000 pounds
加工齿轮上的两个键槽

3.5.4　Serrations Cut Into a Large Shear Blade.

Cutting a special shear blade, see Figure 3-18.

3.5.4　将锯齿状毛坯切削出大的刃口

图 3-18 为切削特殊刃口。

Figure 3-18 Cutting a special shear blade
切削特殊刃口

3.5.5 Splitting Machined Parts

If parts need to be split with traditional machining, the other half is often destroyed. With Wire EDM, the part can be spilt leaving a kerf of just the thickness of the wire. See Figure 3-19.

3.5.5 切割成型部件

如果用传统方式切割成型部件，另一半部件会被损坏。而使用 WEDM 的方式，被加工的部件会仅仅切除小刀的宽度或电极丝的直径宽度，如图 3-19 所示。

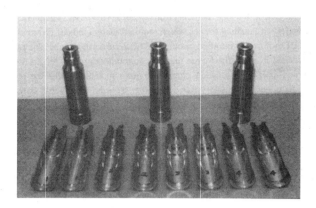

Figure 3-19 Machined parts split in half
将成型部件切割成两半

Questions 习题

1. What is the function of the servo system?
2. Describe the step by step Wire EDM process.

1. 伺服系统的作用是什么？
2. 描述 WEDM 的加工步骤。

3. List advantages of Wire EDM for Die Making. 3. 列举 WEDM 在模具生产中的优点。
4. List three Wire EDM applications. 4. 列出三种 WEDM 适用的场合。
5. List four materials that can be Wire EDMed. 5. 列出四种可以用WEDM加工的材料。
6. What type of material can be cut with Wire EDM? 6. WEDM 可以加工什么种类的材料？

4 Electrochemical Machining 电化学加工

4.1 Fundamentals of ECM 电化学加工基础

Electrochemical machining (ECM) is one of the least used non-traditional machining methods in spite of its high metal removal rate and relatively no electrode wear. This process is used for automotive components, gun barrel rifling, steam turbines, high production machining operations, and for deburring. See Figure 4-1.

电化学加工是最近兴起的特种加工方法之一，它具有高的金属切削率并且相对来说没有电极磨损。主要用来加工汽车零件、炮管膛线、涡轮机和去毛刺等，如图4-1所示。

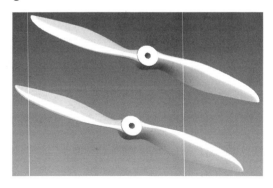

Figure 4-1 Airfoil contours produced with electrochemical machining
利用电化学加工的螺旋桨外轮廓

Electrolysis is the name given to the chemical process which occurs, for example, when an electric current is passed between two electrodes dipped into a liquid solution. A typical example is that of two copper wires connected to a source of direct current and immersed in a solution of copper sulfate in water as shown in Figure 4-2.

电解是一种化学过程，例如这种过程发生在当电流通过浸入电解液的两个电极之间时。一个典型的例子是和直流电源相连的两根铜导线和浸入硫酸铜溶液中的铜电极组成的系统，如图4-2所示。

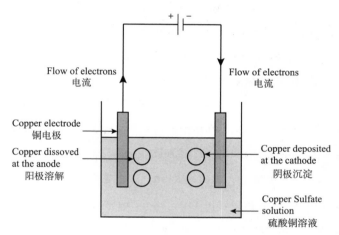

Figure 4-2　Electrochemical cell
电化学电池

An ammeter placed in the circuit, will register the flow of current. From this indication, the electric circuit can be determined to be complete. It is clear that copper sulfate solution obviously has the property that it can conduct electricity. Such a solution is termed as electrolyte. The wires are called electrodes, the one with positive polarity being the anode and the one with

当把一个电流表串联在电路中时就能检测出通过导线的电流。基于这种迹象，我们就可以知道电路是完整的了。硫酸铜溶液有传导电流的能力，这种溶液被称为电解液，两根金属丝则被称为电极，其中一根接电源正极的称为阳极，另一根接电源负极的称为阴极。电极和电解液组成

negative polarity the cathode. The system of electrodes and electrolyte is referred to as the electrolytic cell, while the chemical reactions which occur at the electrodes are called the anodic or cathodic reactions or processes. A typical application of electrolysis is the electroplating and electroforming processes in which metal coatings are deposited upon the surface of a cathode-workpiece. Current densities used are in the order of 10^{-2} to $10^{-1} A/cm^2$ and thickness of the coatings is sometimes more than 1mm. An example of an anodic dissolution operation is electropolishing. Here the workpiece, which is to be polished, is made the anode in an electrolytic cell. Irregularities on its surface are dissolved preferentially so that, on their removal, the surface becomes smooth and polished. A typical current density in this operation would be $10^{-1} A/cm^2$, and polishing is usually achieved on the removal of irregularities as small as 10nm. With both electroplating and electropolishing, the electrolyte is either in motion at low velocities or unstirred.

Electrochemical Machining is similar to electropolishing in that it also is an electrochemical anodic dissolution process in which a direct current with high density and

的系统即称为电解电池，发生在电极上的化学反应称为阳极或阴极反应式过程。电解作用的一个典型应用是电镀和电铸过程，它们都利用阳极溶解的金属沉积在与阴极相连的工件上完成作业。当通入的电流为 $10^{-2} \sim 10^{-1}$ A/cm^2 时，沉积层的厚度有时能超过 1 mm。利用阳极溶解来进行加工的一个典型例子是电解抛光。需要被抛光的工件接在阳极以形成原电池。工件表面不规则的地方首先被溶解，在它们被去除之后，工件表面变得光滑而光亮。在此加工过程中常用的电流密度是 $10^{-1} A/cm^2$，并且抛光过程通常是在去除小至 10 nm 的不规则形状时实现的。对于电镀和电解抛光，所用的电解液通常以低速流动。

电化学加工与电解抛光相似，它也是一个电化学阳极溶解过程，在此过程中，在工件和成型工具（阴极）间通过高密度低

low voltage is passed between a workpiece and a preshaped tool (the cathode). At the anodic workpiece surface, metal is dissolved into metallic ions by the deplating reaction, and thus the tool shape is copied into the workpiece.

The electrolyte is forced to flow through the interelectrode gap with high velocity, usually more than 5m/s, to intensify the mass/charge transfer through the sub layer near anode and to remove the sludge (dissolution products e.g. hydroxide of metal), heat and gas bubbles generated in the gap. In typical manufacturing operations, the tool is fed toward the workpiece while maintaining a small gap. When a potential difference is applied across the electrodes, several possible reactions can occur at the anode and cathode.

Figure 4-3 illustrates the dissolution reaction of iron in sodium chloride (NaCl) water solution as electrolyte. The result of electrolytic dissociation

$$H_2O \longrightarrow H^+ + OH^-$$
$$NaCl \longrightarrow Na^+ + Cl^-$$

are negatively charged anions: OH^- and Cl^- towards to anode, and positively charged cations: H^+ and Na^+ towards to cathode.

At the anode:

$$Fe \longrightarrow Fe^{2+} + 2e$$

电压直流电。在阳极工件的表面，金属通过退镀过程被溶解成阳离子，因此工件就能被复制成工具的形状。

为了扩大通过阳极附近分层之间的质量和电荷转移并且可以去除沉积的油泥（例如溶解产物氢氧化金属）以及在缝隙间产生的热量和气泡，在电极之间总是通以高于 5 m/s 的电解液。在典型的加工过程中，刀具在保持小间隙的同时向工件进给。当电极之间施加电位差时，阴阳极之间还可能会发生几种反应。

图 4-3 显示了当以氯化钠溶液作为电解液时的溶解反应。其电离方程式如下：

其电解的结果是其中带负电荷的阴离子：氢氧根离子和氯离子朝阳极运动；而带正电荷的阳离子：氢离子和钠离子则朝阴极运动。

阳极反应：

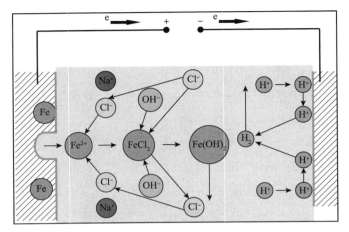

Figure 4-3 Diagram of electrochemical reactions during ECM of iron in sodium chloride (NaCl) electrolyte
当采用氯化钠溶液作为电解液时在电化学加工过程中发生的电化学反应

At the cathode, the reaction is likely to be generation of hydrogen gas and the hydroxyl ions:

$$2H_2O + 2e \longrightarrow H_2 + 2OH^-$$

The outcome of these electrochemical reactions is that the iron ions combine with other ions to precipitate out as iron hydroxide Fe(OH)$_2$. The ferrous hydroxide may react further with water and oxygen to form ferric hydroxide:

$$4Fe(OH)_2 + 2H_2O + O_2 \longrightarrow 4Fe(OH)_3$$

although it is stressed that this reaction, too, does not form part of the electrolysis. The salt (for example, NaCl) is not consumed in the electrochemical processes, therefore, for

而在阴极，反应很可能产生氢气和氢氧根离子：

这些电化学反应的最终结果就是铁离子和其他的离子结合最终以氢氧化亚铁的形式存在。而氢氧化亚铁可能会进一步和水及氧气反应形成氢氧化铁：

尽管这个反应会剧烈地进行，但是它并不是电解的一部分。电解液中的盐（如氯化钠）在电化学加工时并没有被消耗，

keeping constant concentration of electrolyte, it may be necessary to add more water.

With this metal-electrolyte combination, the electrolysis has involved the dissolution of iron from the anode, and the generation of hydrogen at the cathode. No other actions take place at the electrodes. Electrochemical Machining is a relatively new and important method of removing metal by anodic dissolution and offers a number of advantages over other machining methods. Metal removal is effected by a suitably shaped tool electrode, and the parts thus produced have the specified shape, dimensions, and surface finish. ECM forming is carried out so that the shape of the tool electrode is transferred onto, or duplicated in, the workpiece. For high accuracy in shape duplication and high rates of metal removal, the process is effected at very high current densities of the order $10 \sim 100 \text{A/cm}^2$, at relative low voltage usually from 8 to 30V, while maintaining a very narrow machining gap (of the order of 0.1mm) by feeding the tool electrode in the direction of metal removal from the work surface, with feed rate from 0.1 to 20mm/min. Dissolved material, gas, and heat are removed from the narrow machining gap by the flow of electrolyte pumped through the gap at a high

因此为了保持电解液的浓度，需往电解液中加入更多的水。

在电解过程和金属电解液的共同作用下，金属阳离子不断从阳极溶解，而在阴极不断产生氢气。在电极上没有其他的反应发生。电化学加工是一种相对比较新颖、利用阳极溶解来去除材料的加工方法，相比其他加工方法有很多优点。金属的去除是通过适当形状的工具电极来实现的，因此加工的产品具有指定的形状、尺寸以及表面粗糙度。电化学成型最后将工具电极的形状复制到了工件上。为了获得更高的形状复制精度和更高的金属去除效率，加工过程需要的电流密度为 $10 \sim 100 \text{ A/cm}^2$，电压相对较低，为 $8 \sim 30 \text{ V}$。为了维持工具和工件之间的微小加工间隙（通常要求为 0.1 mm）以保证工具电极能从垂直于工件表面的方向进给切除材料，进给量一般为 0.1～20 mm/min。高速（5~50 m/s）电解液将会从加工间隙中去除被切除的材料、产生的气体以及热量。作为一种不需要机械能来去除材料的加工过程，电化学加工可以以高效率来加工导电材料。

velocity (5~50m/s). Being a non-mechanical metal removal process, ECM is capable of machining any electricallyconductive material with high stock removal rates regardless of their mechanical properties. In particular, removal rate in ECM is independent of the hardness, toughness and other properties of the material being machined. The use of ECM is most warranted in the manufacture of complex-shaped parts from materials that lend themselves poorly to machining by other, above all mechanical, methods. There is no need to use a tool made of a harder material than the workpiece, and there is practically no tool wear. Since there is no contact between the tool and the work, ECM is the machining method of choice in the case of thin-walled, easily deformable components and also brittle materials likely to develop cracks in the surface layer.

As mentioned above, in most modifications of ECM, the shape of the tool electrode is duplicated over the entire surface of the workpiece connected as the anode. Therefore, complex-shaped parts can be produced by simply moving the tool translationally. For this reason and also because ECM leaves no burrs, one ECM operation can replace several operations of mechanical machining. ECM

尤其需要指出的是，电化学加工的效率与被加工材料的硬度、刚度和其他属性无关。电化学加工已经成为加工复杂形状工件的首选加工方法，这些工件的材料本身不适合用其他机械方法加工。它不需要用比工件硬度高的工具来加工，而且几乎没有工具的磨损，这是因为工具和工件之间根本就没有接触。电化学加工也成了薄壁且容易变形等易在表层发生破裂的脆性材料的首选加工方法。

正如上面所提到的，在绝大多数电化学加工过程中，工具电极的形状可以完全被复制到与阳极相连的工件上。因此，复杂形状可以比较容易地从工具电极上转移过来。而且电化学加工不会留下毛刺，一道电化学工序可以代替多道机械加工工序。电化学加工可以去除材料不完整的表

removes the defective layer of the material and eliminates the flaws inherited by the surface layer from a previous treatment and usually no generated residual stress in the workpiece. All this enhances the service qualities of the parts manufactured by ECM. Simultaneously, ECM suffers from several drawbacks. Above all, it is not at all easy to duplicate the shape of the tool electrode in the workpiece to a high degree of accuracy because there is some difficulty in confining the ECM process precisely within the areas that must be machined. At this writing, a fairly consistent theory has been formulated to explain the anodic dissolution of metals, alloys and composites, and mathematical techniques describing the ECM process and computer simulations have been advanced for the design of tool electrodes and process parameter control. Recent years have seen the emergence of ECM manufacturing centers and computer-aided system to design tool electrodes.

Some metal is also dissolved from adjacent areas on the workpiece. ECM machines are often equal, if not more, expensive than conventional metal-cutting machines and need more floor area for their installation. The electrolytes used in ECM

面，还可以消除由于先前的处理而产生的裂纹并且不会产生应力集中现象。这些全部都是电化学加工在提高工件加工质量方面的优点。同时，电化学加工也存在着以下一些缺点。首先，要高精度地轻松复制工具电极的形状并不是一件易事，因为要确保电化学加工过程精确进行，那么被复制的工具内部表面就必须被加工处理过。在这方面，已经形成了一套连续理论来解释金属、合金及各组成的阳极溶解过程，为电化学加工提供了机械加工技术以及可以提高工具电极设计和过程参数控制的计算机仿真软件等。近年来，还出现了电化学加工中心以及计算机辅助工具电极设计系统等。

工件上的部分金属也会在加工中心周围溶解。电化学加工设备相比传统金属切割设备一直很普通且便宜，并且不需要占用更多的地方来安装。电化学加工使用的电解液会腐蚀设备。有关产

attack the equipment. The environmental problems, which are connected with the utilization of generated waste, are very important in ECM. Various methods have been proposed to recover and re-use ECM sludge.

生废物的利用和环保问题是电化学加工的一个重要研究问题。目前，在如何恢复和再利用电化学沉积物的问题上形成了多种理论并进行了应用。

4.2　How does Electrochemical Machining work
　　 电化学加工过程

Electrochemical machining is similar to EDM, since it uses a shaped electrode, to machine the workpiece. With EDM, the workpiece is submerged in a tank of dielectric fluid as electrical discharges from an electrode erode material from the workpiece. With electrochemical machining, a pressurized conductive salt solution, the electrolyte, flows around the workpiece as an electrical-chemical reaction deplates material from the workpiece. See Figure 4-4 for the electrochemical process.

电化学加工类似于电火花加工，因为它们都利用成型电极来加工工件。在电火花加工时，通常是将工件淹没在绝缘液体中以此来保证从电极放出的电流能够侵蚀到工件材料。而电化学加工时，通常用带压导电盐溶液作为电解溶液使其流过工件四周来发生材料的电化学反应，图4-4是电化学加工过程。

As the electrolyte flows between the narrow gap of the conductive workpiece and the shaped electrode, an electrical current of low DC voltage and high amperage is applied. The current is insufficient to produce a spark; instead, the metal dissolves from the workpiece by electrochemical reaction

当电解液在成型电极和导电工件之间流通时，施加低电压高安培的电流。电流将不会产生火花，金属将通过电化学反应被融解而工具电极将不会有任何明显的磨损。当提供的电流合适且金属工件中的原子以离子形式进入

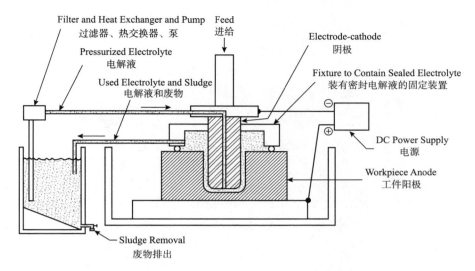

Figure 4-4　The electrochemical process
电化学加工过程

without any noticeable electrode wear. The controlled deplating occurs as electrical current is applied and the metal ion of the workpiece removes to an ion in the electrolyte solution.

The process of electrochemical machining is the opposite of plating. With plating, metal is applied with the use of electrical current; with electrochemical machining, metal is removed with the use of electrical current. The electrolyte, pumped at high velocity between the electrode and the workpiece, removes the dissolved metal and heat. The electrolyte is then pumped into a tank where the sludge is eliminated, and a heat exchanger removes the heat. The filtered cooled electrolyte is then reused. See Figure 4-5 for

电解液中时，则发生退镀过程。

电化学加工过程是电镀过程的逆过程。电镀过程是利用电流来"敷"一层金属，而电化学加工是利用电流来去除金属。在电极和工具之间泵送高速电解液以带走熔化的金属和产生的热量。循环的电解液将会被泵带入一个容器内并且该容器会将产生的废物过滤干净，产生的热量也通过热交换器被清除，在此过程中，过滤器还可以被反复用来冷却电解液。图 4-5 所示的是电化学加

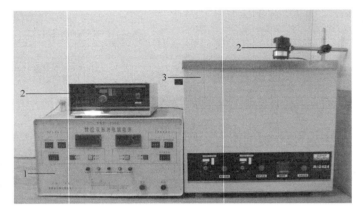

Figure 4-5　An electrochemical machine
1- Plus power, 2-Electronic churn device, 3-Ultrasonic generator

电化学加工机器
1—脉冲电源；2—电子搅拌器；3—超声波产生器

an electrochemical machine.

Basic operating parameters of ECM are:

(1) Working voltage between the tool electrode (cathode) and the workpiece (anode).

(2) Machining feed rate.

(3) Inlet and outlet pressure of electrolyte (or flow rate).

(4) Inlet temperature of electrolyte

The value of current used in ECM is dependent on the above parameters and dimensions of the machining surface. For manufacturing results of ECM, the distribution of current density on the anode surface and the distribution of gap size between the electrodes are very important, which depend on above parameters and electrochemical properties of

工机器。

电化学加工的基本参数：

（1）电极（阴极）和工具（阳极）之间的工作电压；

（2）加工进给率；

（3）电解液的入口和出口压力；

（4）电解液的入口温度。

电化学加工中使用的电流主要根据以上参数和加工表面尺寸来选择。为了获得较高的加工质量，在阳极表面的电流密度分配和加工间隙间的电流密度分配显得十分重要。而这种分配的主要依据是以上的参数和工件的电化学性能以及电解液的组成及性

workpiece material and electrolyte. Typical values of parameters and conditions of ECM are presented in the Table 4-1.

能。常用典型参数值和电化学加工条件如表 4-1 所示。

Table 4-1 Typical values of parameters and conditions of ECM
电化学典型参数值和加工条件

Type 类型	Direct Current 直流
Voltage 电压	5 to 30 V (continue or pulse) 5~30 V（连续或脉冲）
Current 电流	50 to 40,000 A 5~40000 A
Current Density 电流密度	10 to 500 A/cm² 10~500 A/cm²
Temperature 温度	20 to 60 ℃ 20~60 ℃
Feed rate 进给量	0.1 to 20mm/min 0.1~20 mm/min
Electrode material 电极材料	Brass, copper, bronze 黄铜、铜、青铜

4.3 Advantages of Electrochemical Machining 电化学加工的优点

4.3.1 Practically No Electrode Wear

A great advantage of the electrochemical machining process it produces practically no wear on the electrode. Thousands of parts can be made with just one electrode.

4.3.2 No Recast Layer or Thermal Stress

With EDM, there is a recast layer; however, with electrochemical machining

4.3.1 几乎没有电极磨损

电化学加工最大的优点是加工制造时几乎没有电极磨损。上千件产品甚至可以只用一个电极来加工。

4.3.2 不产生再铸层或热应力

电火花加工时会产生一个再铸层，而电化学加工却不存在再

there is no recast. EDM uses spark erosion to remove metal. Electrochemical machining dissolves the metal and produces no thermal stress.

4.3.3 Material Hardness and Toughness Not a Factor

Electrochemical machining removes metal atom by atom; the material hardness and toughness has little effect on this process.

4.3.4 Rapid Metal Removal

Electrochemical machining removes metal rapidly. A large electrochemical machining job shop applies this rule: with 10,000amps of power-one cubic inch of metal removal per minute. This particular job shop reports that the jet engine business provides 99% of its work. Although electrodes are difficult to produce, the company says that it produces jet engine air foils holding ±0.051mm.

4.3.5 Deburring and Radiusing of Holes

An ideal use of ECM is the deburring and radiusing of holes as illustrated in Figure 4-6.

Figure 4-7 shows what happens when the electrode in electrochemical machining remains stationary in relation to the workpiece. The current flows between the electrode and the workpiece, but the

铸层。电火花加工利用火花腐蚀去除金属材料,而电化学加工则通过溶解金属来作用且不会产生热应力。

4.3.3 材料的硬度和刚度不再受限制

电化学加工通过原子来去除金属原子,而材料的硬度和刚度对加工过程没有影响。

4.3.4 较高的材料去除率

电化学加工去除材料的速度很快,很多电化学加工厂普遍采用以下原则:每分钟去除每立方英尺的金属使用10000A的能量。据报道,喷气式发动机生产厂家99%使用电化学加工。尽管电极很难制作,但是制造商表示,这种喷气式引擎箔片加工甚至能达到±0.051 mm的精度。

4.3.5 修毛刺和倒圆角

在给孔修毛刺和倒圆角时利用电化学加工能得到非常理想的效果,如图4-6所示。

图4-7所示的是电化学加工时电极和工件是如何保持相对静止的。电流在电极和工件之间流过,但是加工只发生在工件表面,以此来控制修毛刺和倒圆角

Figure 4-6　Various parts that have been deburred with electrochemical machining
　　　　　利用电化学加工修毛刺后的工件

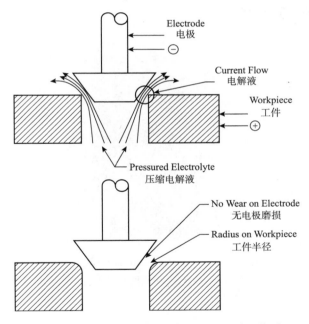

Figure 4-7　Electrochemical deburring and radiusing
　　　　　利用电化学加工修毛刺和倒圆角

machining takes place only on the workpiece. This controlled machining results in deburred or radiused holes.

One company uses the electrochemical

的加工效果。

某企业利用电化学加工来给

machining process to deburr and radius holes for automotive air bag inflator housings. Typically, these housings contain fifty holes that allow gas to inflate the air bags within milliseconds. The hole edges are critical to the performance of these units. ECM can deburr and put radiuses on eight parts in less than ten seconds. The fully automated machine loads, unloads, and washes the air bag inflator housings—it produces 20,000 parts a day. See Figure 4-8.

汽车气囊充气装置修毛刺和倒圆角。这些装置上设计了50个孔，以此来保证气体能在几毫秒内被迅速压入气囊内。这些孔的边缘质量将直接影响到气囊装置的性能。而电化学加工则可以在不到10秒内给8个这样的工件修毛刺并倒圆角。在全自动装载、卸载和清洗气囊充气装置生产线上，利用电化学加工一天可以加工20000个工件，如图4-8所示。

Figure 4-8　Deburring and radiusing of holes
修毛刺和倒圆角之后的产品

4.4　Disadvantages of Electrochemical Machining 电化学加工的缺点

4.4.1　The Shaped Workpiece is Not a Replica of the Electrode
4.4.1　成型工件并不完全是电极的复制品

The flow of the electrolyte and the 电解液的流动和电极的形状

shape of the electrode are critical to the shape that will be made in the workpiece. In other words, a square electrode will not necessarily produce a square form with electrochemical machining. The electrolyte flowing around the electrode determines the produced shape. With EDM, the shape produced in the workpiece is a replica of the electrode.

4.4.2 Shaped Electrodes are Difficult to Machine

The electrochemical machining process requires skilled craftspersons to produce electrodes, because the electrodes are not replicas of the workpieces. This process is generally used only for moderate to high production items.

4.4.3 Electrolyte and Sludge Removal

The highly corrosive electrolyte is a salt solution with additives, also the electrochemical machining process produces metal hydroxides, or hydrates, that become sludge. Filters, settling, or centrifugal devices are used to remove the sludge from the electrolyte. The amount of sludge removal can be 100 to 500 times the volume of the material removed. The sludge can be put into a filter press where it is condensed to 40 to 50 percent solid matter, or the metal can be removed from the sludge. This sludge removal and

对于工件的成型至关重要。换句话说，在电化学加工时，方电极不一定会产生一个方形。在电极周围流动的电解液决定了加工的形状。而利用电火花加工，工件成型后完全是电极的复制品。

4.4.2 成型电极加工困难

电化学加工过程需要工艺精湛的工匠来加工电极，因为电极并不完全是工件的复制品。这个过程通常只用于中高产量的项目。

4.4.3 电解液和油污的清理

具有强腐蚀性的电解液是由盐溶液和一些添加剂组成，并且在电化学加工过程中还产生金属氧化物或氢氧化物变成油污。过滤器和沉淀装置以及离心装置都是用来去除电解液中的油污，在净化油污的系统内可能会对油污进行 100~500 次的净化。油污可以被压入过滤器中并被过滤过 40%~50% 的固体颗粒。由于这种油污对于环境是一种极大地破坏，一个大型的电化学加工厂必

treatment process is an important factor when considering electrochemical machining. Because sludge can be hazardous to the environment, one large electrochemical machining company removes all the metal from the sludge. This company produces from four to five tons of metal a week from the sludge.

须将油污中的金属全部回收，而这样的工厂一周就可以从油污中回收 4~5t 金属废物。

4.5 Pulse Electrochemical Machining　脉冲电化学加工

ECM practice leads to the conclusion that machining inaccuracy is proportional to the gap size. The shape error is dependent on deviation of properties in the gap medium and physical conditions, such as electrical conductivity, temperature, void fraction, current efficiency (electrochemical machinability), flow velocity, pressure etc.

Therefore, for improvement of shape accuracy and simplification of tool design, the gap size during ECM should be as small as possible. Additionally, more stabilization of the gap state is needed by reducing non-uniformity of electrical conductivity and other physical conditions, which are significant for the dissolution process.

All these requirements for ECM performance, with continuous working

在实际的电化学加工中往往会得到这样一个结论，即加工中的误差与间隙大小成比例。形状误差主要取决于在间隙之间的媒介和物理条件。例如电传导率、温度、空隙率、电流效率、流体速度、压力等。

因此，为了提高复制形状的精度并简化工具设计，在电化学加工过程中，间隙的大小应尽可能小。此外，为了获得间隙状态的稳定就需要减小电传导率和其他物理条件的不均匀性，这对于溶解过程是非常重要的。

所有的这些电化学加工要求在持续的工作电压下都是有限

voltage, is very limited. The minimum practical tool gap size, which may be employed, however is constrained by the onset of unwanted electrical discharges. These short electrical circuits reduce the surface quality of the workpiece, and led to electro-erosive wear of the tool-electrode, and usually machining cannot progress because of them. Investigations of electrical discharges in an electrolyte reveal that the probability of electrical breakdown in the gap is a function of the evolution of gaseous-vapor layers and passivation of the work surface. Intense heating, hydrogen generation sometimes choking phenomena and cavitation within the gap can lead to evaporation and subsequent gas evolution, and it is this gas which is believed to cause the onset of electrical discharge. The issue of heating of electrolyte is of primary importance for the determination of limit conditions of ECM process. The distribution of mean temperature in the inter-electrode gap along the flow was determined using one-dimensional mathematical model of ECM process. Further specification of temperature distribution was revealed. Due to the heat exchange through electrodes as well as distribution of electrode, the temperature changes along the flow path

的。然而，可以使用的最小的实际加工间隙受到不必要的放电开始的限制。这些短暂的电流回路会影响工件表面的加工质量，并会造成工具电极的电腐蚀，而且常常由于它们的存在，加工不能进行。通过对在一种电解液中的放电现象的研究揭示出电流会成为汽化层演变和工件表面钝化等破坏的主要可能原因。过热、氢气的产生有时会产生堵塞和在间隙中产生气穴等现象，而这可能会导致蒸发和随后进一步的汽化过程，就是因为这个过程产生的气体才最终引起初始的放电现象。电解液中的热量也成为电化学加工中非常重要的限制条件。这种在电解液中的电极内部间隙的热量分配可以用一个一维数学模型来阐释，而温度分配的特殊性也被揭示出来。由于通过电极的热量交换和电极间的热量分配，通过间隙来散热的同时也沿着电解液通道散热。

as well as across the gap size.

In the pulse ECM process, a pulse generator is used to supply the working voltage pulses across the two electrodes, typically in the form of pulse strings consisting of single pulses or grouped pulses. See Figure 4-9.

在脉冲电化学加工过程中，脉冲发生器用在两电极之间来提供脉冲工作电压。通常是由单脉冲或多脉冲组成的连续脉冲形式，如图4-9所示。

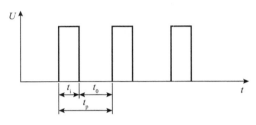

Figure 4-9　Plus strings
连续脉冲

The anodic electrochemical dissolution occurs during the short pulse on-times, each raging from 0.1 to 5ms. Dissolution products (sludge, gas bubbles and heat) can be flushed away from the inter-electrode gap by the flowing electrolyte during the pulse off-times between two pulses or two groups of pulses. To intensify the electrolyte flushing, the tool is retracted from the workpiece to enlarge the gap during the pulse off-times. The gap checking and tool repositioning can also be conducted during these pulse pauses to establish a given gap size before the arrival of the next pulse, leading to a significant reduction in the indeterminacy of the gap

阳极电化学融解发生在短脉冲宽度时，每次在0.1~5 ms变化。电极间隙的溶解产物（金属废物、气体和热量）可以在两脉冲或两组脉冲间隔内被电解液移除。为了增强电解液冲刷的效果，在脉冲间隔内可以将工具从工件上撤回来增大间隙。对间隙的控制和工具的重新定位可以在这些脉冲停止的时候来进行，以便在下一次脉冲发生时建立一个给定的间隙大小。这样就可以大幅度地减小间隙和加工精度的不确定性。在精密电化学加工持续作业中可以看到在尺寸控制、造

and, hence, of the shaping accuracy. Existing work on PECM has shown considerable improvements in dimensional controllability, shaping accuracy, process stability, and simplification of tool design. With PECM, it is possible to reproduce complex shapes, such as dies, turbine blades, and precision electronic components, with accuracy within 0.02 to 0.10mm. From theory and practice of PECM it follows that the pulse-off time should be long enough, firstly, for the gas and heat to be removed from the gap, secondly, for the metal to revert from activated to the passive state. As for the second point, to this end it was proposed to apply small pulses of the reverse polarity between the high-power working pulses. To eliminate the variations in the medium conductivity along the interelectrode gap in the direction of electrolyte flow, the larger the distance between the gap inlet and outlet, the longer should be the pulse-off time. In the machining of large-sized parts, the pulse-off time should be so long that the ECM productivity becomes too low. To overcome this drawback of pulsed ECM, special pulsed-cyclic regimes were elaborated, where the processes of metal removal and washing of inter-electrode gap are separated in

型精度、过程稳定和简化工具设计方面的巨大改进。在精密电化学加工中，重塑复杂形状成为可能，例如冲模、涡轮叶片和一些精密电子元件等，利用精密电化学加工可以达到 0.02 ~ 0.10 mm 的精度。从精密电化学理论和实践中发现脉冲间隔需要控制得足够长。首先，为了使产生的气体和热量从间隙中排出；其次，为了使金属从活跃状态转变为稳定状态。关于第二点，为了终止它，建议在高能工作脉冲之间使用相反极性的小脉冲。为了消除在电解液流动方向沿着极间间隙的电导率的变动率影响，进口和出口间的距离越大，则脉冲间隔时间应越长。在大型工件的加工中，脉冲间隔时间过长会大大地降低电化学加工效率。为了克服这个缺点，就形成了一种特殊脉冲循环模式，在此模式中去除金属和极间间隙的清洗两过程在不同的时段进行。这个在 TE 和 WP 间的时间段根据一个给定的关系来变化。脉冲电压随着电极的移动而同时加载。而极间距离很小时，加工时通常使用多脉冲电流，而间隙间电解液的流动率

time. The distance between the TE and WP varies according to a given relationship. A voltage pulse is applied in synchronism with the electrode movement. When the inter-electrode distance is small, the machining is performed with using usually the group pulses. The electrolyte flow rate in this small gap is usually low. When the distance is large, the gap is washed with intensive electrolyte flow. Under these conditions (a large gap), washing may be performed rapidly and the pulse-off time should not be too long. In this case, it becomes possible to perform ECM with very small gaps (several hundredths parts of a millimeter) enabling one to gain high accuracy of ECM.

The term "Pulse Electrochemical Machining" includes several modifications of realization of discrete electrochemical shaping. Existing methods and schemes of PECM depend on a number of controlled factors, and they can be divided into the following groups:

(1) According to kinematics of the tool electrode and the workpiece.

(2) According to control systems.

(3) According to shape of pulses and its distribution in time.

总是很低。当极间距离较大时，间隙可以被大量的电解液冲刷，在这种大间隙条件下，冲刷可以很快完成且脉冲间隔时间不会太长。在这种情况下，就可以进行小间隙的电化学加工来获得高精度的电化学产品。

术语"脉冲电化学加工"包含几个分离的电化学加工工艺手段。而精密电化学加工理论依靠许多控制因素，主要分为以下几条：

（1）依靠工具电极和工件的运动学关系；

（2）依靠控制系统；

（3）依靠脉冲的形状及其分配。

4.6 ECM Applications　电化学加工的应用

Few companies use electrochemical machining, in spite of its advantages of high metal removal rate, no electrode wear, and ability to cut difficult metals. The major difficulties for electrochemical machining are:

(1)The highly corrosive effects of the electrolyte.

(2)The electrode not being a replica of the workpiece.

(3)The difficulties in producing the electrodes and the tooling.

In spite of electrochemical machining drawbacks, there are useful applications for this process. Electrochemical machining can be profitable for parts requiring high surface quality, high production, and when other machining methods are inefficient. ECM applications are listed as follows:

4.6.1　Stem Drilling

The stem drilling process drills small deep holes. Electrolyte is pumped through a hollow acid resistant tube, usually made of titanium. Except for the tip, the entire tube is insulated. As the tip approaches the workpiece and the electrolyte flows from

很少有企业使用电化学加工，尽管它有很多优点，例如高的金属切削率、无电极磨损、可加工不易加工金属等。电化学加工的主要困难是：

（1）电解液具有过高的腐蚀性；

（2）加工后的工件与电极形状相差甚远；

（3）需要用的工具和电极加工困难。

尽管电化学加工存在以上缺点，但仍有很多应用的地方。电化学加工常用于加工要求高表面质量、高产量和其他加工手段无法加工的工件。电化学加工常见应用如下。

4.6.1　茎状孔加工

电化学加工可以用来加工小而深的茎状孔。电解液通过一个中空的抗酸钛管泵入，除了尖端，整个管子都是绝缘的。当管尖端接触到工件而电解液从管子中心被泵入时，工件就发

its center, the workpiece erodes. The used electrolyte flows past the insulated tube and the walls of the hole. Often the electrolyte is filtered and reused. Various sized and shaped electrodes can be used at the same time. See Figure 4-10 and Figure 4-11.

4.6.2 Capillary Drilling

Capillary drilling is similar to stem drilling except that the tube consists of glass with a metal wire running through the tube center. As the electrolyte flows through the

生腐蚀。而电解液则冲刷了绝缘管和孔壁,并被过滤重新使用。不同尺寸和形状的电极可以同时使用,如图4-10和图4-11所示。

4.6.2 毛细孔加工

毛细孔加工类似于茎状孔加工。但所使用的管子是中心穿过一根金属线的玻璃管。当电解液泵入精细的玻璃管并沿着金属

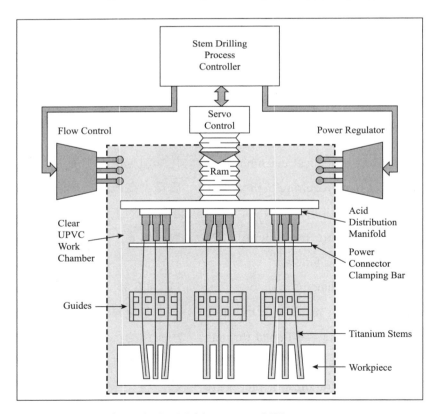

Figure 4-10　Multi-part stem drilling system
批量工件的茎状孔加工系统

Figure 4-11　Gas turbine vane with 54 trailing edge holes drilled in one operation
汽轮机叶片上 54 个孔一次加工成型

fine glass tube and electricity flows through the wire, the workpiece erodes at the tip. This process is used when the hole diameters are under 0.4mm and the depths exceeds the ratio of 10 to 1.

4.6.3　Preparing Nanocomposite Coatings

Nanocomposite coatings can be prepared by several techniques, such as sputtering, ion implantation and electrodeposition. The electrodeposition process is relatively easy to control and can be used to incorporate a variety of ceramic particles into different metals, and thus is widely used. Figure 4-12 shows the atomic force microscopy (AFM) images of Ni-TiN nanocomposite coating.

线流入时，工件就在尖端发生腐蚀。这种加工通常用于直径小于 0.4 mm 而深度超过 10 倍直径的孔的加工。

4.6.3　纳米复合材料涂层的预处理

纳米复合材料涂层可以用多种技术来做预处理，例如喷溅、离子注入和电极沉积技术。电极沉积过程相对容易控制并且可以被用来往不同的金属中注入多种陶瓷材料。这种技术有着广泛的应用。图 4-12 所示的是 Ni-TiN 纳米复合材料涂层的原子显微镜成像。

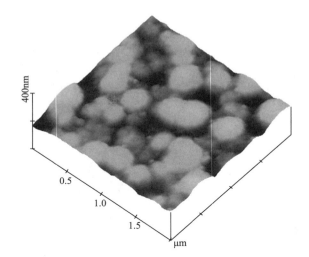

Figure 4-12　AFM images of Ni-TiN nanocomposite coating.
Ni-TiN 纳米复合材料涂层的原子显微镜成像

Questions　习题

1. What will happen when two copper wires are connected to a source of direct current and immersed in a solution of copper sulfate in water?

2. Describe the electrochemical machining process.

3. What are the advantages of electrochemical machining?

4. What are the disadvantages of electrochemical machining?

5. Why do we have to use the pulse electrochemical machining?

6. List some applications of electrochemical machining.

1. 如果将两根铜导线连上直流电源并浸入硫酸铜溶液中会发生什么现象？

2. 描述电化学加工的过程。

3. 电化学加工的优点是什么？

4. 电化学加工的缺点是什么？

5. 为什么要使用脉冲电化学加工？

6. 列举电化学加工的应用。

Lasers Machining 激光加工 5

5.1 Fundamentals of Lasers　激光加工基础

Lasers can be found in all sorts of human endeavors. Lasers drill diamonds, teeth, and steel. Lasers weld computer chips, autos, and detached retinas. Trained eye surgeons strike the retina with a blue argon laser beam which bonds the detached retina. Before this breakthrough technology, traumatic freezing and surgical procedures were used.

Lasers align machines, read merchandise codes, print documents, play audio discs, and send images and sound over thin fiber optic cables thousands of miles. Coherent laser light beams can be as high as one million times more powerful than earth's sun ray beams. The following chapters on lasers will include cutting, welding, cladding, drilling,

激光加工可以在人类的诸多活动里找到踪迹，例如用激光给钻石、牙齿、钢材打孔，用激光焊接计算机芯片、汽车和分离的视网膜。训练有素的眼科医生会使用蓝色激光束来将两块分离的视网膜连接成一体。在这项突破性技术被发明之前，一般是使用创伤性冷冻和外科手术。

激光可以校准机器、读取商品条形码、打印文件、播放歌曲，并且能够通过光缆从数千英里之外传送图片或声音信息。相干激光光束的能量是太阳照射到地球表面光束的一百万倍。以下章节将介绍激光切割、激光焊接、激光熔覆、激光钻孔、激光

hardening, marking, and other procedures related to laser manufacturing.

5.1.1 Laser Cutting

Increasingly, laser cutting is becoming the first choice for many manufacturers. Laser cutting machines range in power to 5000 watts and cut with speeds from 25 to 15,000mm per minute. See Figure 5-1.

5.1.1 激光切削

如今，激光切削已经成为许多加工者的首选加工方法。激光加工的功率可达 5000 W，其切削速度为 25～15000 mm/min，如图 5-1 所示。

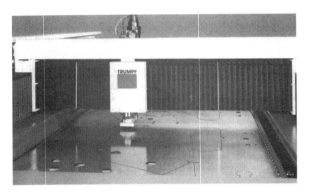

Figure 5-1　A CO_2 cutting laser
CO_2 激光切割器

The accuracy, speed, and minimum workpiece distortion of lasers have altered the manufacturing of many parts. On a worldwide basis for industrial lasers, over 40% of the lasers are used for cutting, about 20% for welding, and the remainder for drilling, hardening, cladding, marking, and others.

In 1960, Ted Maiman of Hughes Corporation developed the first laser from a ruby crystal. The intensity of the light

激光加工的精度、速度、最小变形量已经改变了许多工件的加工。在世界上所有激光加工的应用中，超过 40% 的激光加工用于切割，约 20% 用于焊接，其余的应用于打孔、硬化、激光镀、打标识等。

1960 年，休斯公司的 Ted Maiman 发明了第一台红宝石晶体激光加工机，激光的强度超过

surpassed anything known before that time. The first production laser was installed in 1965 when a ruby laser drilled small holes in diamonds. Today, thousands of lasers perform various operations in manufacturing and job shops around the world.

Lasers are extremely cost effective because they eliminate expensive hard tooling. Lasers can economically produce round holes, square cutouts, radii, tapers, and undercuts to any imaginable shape, without expensive tooling costs as with turret or power presses. Looking at the parts in Figure 5-2, one could imagine the expensive tooling that would be needed if lasers or other cutting systems were unavailable.

了当时已知的所有东西。激光加工的首次应用是在 1965 年，当时使用的是红宝石激光切割器，应用于在钻石上打孔。今天，成千上万的激光加工方式在世界加工工业中扮演着不可或缺的角色。

激光加工非常经济，因为它消除了昂贵的硬质刀具。由于不必在硬质刀具方面过多花费，激光加工可以极其经济地加工圆孔、面、圆弧、锥面，甚至任何可以想象到的形状。如图 5-2 所示的零件，我们可以想象，若没有激光加工，我们将会在刀具上浪费多少资金。

Figure 5-2 Various shaped parts cut with lasers
使用激光加工切割出的各种形状的零件

To produce the various shapes, a desired shape is programmed on a CAD system, post processed to convert CAD design language to CNC language, and then downloaded into a

为了生产各种各样的形状，我们常常使用 CAD 程序画出想要的形状，并将 CAD 语言编译成为 CNC 语言，并导入激光切

laser. In most cases, the laser cut part requires no further finishing. Understanding lasers can result in substantial increases in productivity. See Figure 5-3.

割机的 CNC 系统中。在绝大多数情况下，激光加工不需要精加工。使用激光加工的手段可以大大地提高生产率，如图5-3所示。

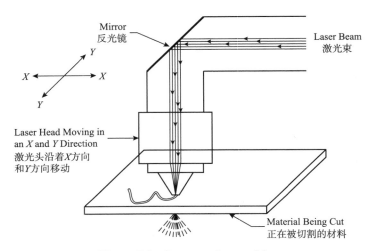

Figure 5-3　Laser cutting machine
激光切割机

5.1.2　How Lasers Work

Laser, an acronym for Light Amplification by Stimulated Emission of Radiation, receives its energy from the amplification of light. A laser consists of a lasing medium such as gas, crystal or liquid; an energy source to stimulate an emission of photons from the medium such as electricity, flashlamp, or another laser; and mirrors to provide optical feedback and amplification. Some of the light energy is allowed to escape at one end of the laser, and then the beam can be directed by mirrors or fiber optics to the workpiece.

5.1.2　激光加工过程

激光加工是 Light Amplification By Stimulated Emission of Radiation 的缩写，其能量源于光能的放大化。一台激光加工器一般包括激光介质（如气体、水晶或者液体）、刺激光量子从介质中散发的能源（如电能、闪光灯或其他激光切割机）、提供反射和光能扩大化的反光镜。有些光能会溢出激光发生器，然后光束将被反光镜或光纤导向工件表面。

5.1.3 Speed of Lasers

The cutting speed of lasers is determined by the type of material, the material thickness, the assist gas, beam quality, and the required edge finish. Generally, the material type and thickness cannot be changed. For non-critical parts, allowing for a coarser edge surface allows the laser to cut faster.

Generally, oxygen cuts stainless steel faster; however, it leaves a dark edge with an oxide scale. This dark edge hinders welding. Though more costly, nitrogen can be used to cut stainless steel to produce a clean edge for welding.

5.1.4 Tolerances

The closer the tolerances, the slower the cutting speeds. On thin work, lasers can hold up to ±0.08mm; some claim ±0.025mm and closer. However, whenever possible, more tolerances should be allowed, generally ±0.13mm.

Many materials have internal stresses. When these materials are cut with laser, the parts often move during the cutting process. This factor should be considered when

5.1.3 激光加工的加工速度

激光加工的速度取决于其加工的材料属性、材料的厚度、辅助气体、激光束质量以及加工边界的粗糙度。通常情况下，材料类型和厚度是无法改变的。对于非关键零部件，允许其加工的边界稍有粗糙，激光切割速度更快一些。

通常，切割不锈钢的时候，氧气作为辅助气体时切削速度较快，但是加工完毕时会留下一条氧化的黑色边界。这条黑色边界不利于焊接，所以尽管氮气成本更高，我们还是在加工不锈钢时使用氮气作为保护气体以提供一个洁净的焊接边界。

5.1.4 公差等级

公差等级越高，加工速度就越慢。在薄件加工时，激光加工可以保持 ±0.08 mm 的形状误差；有些甚至可以达到 ±0.025 mm 或是更小的公差值。然而，在允许的情况下，我们倾向于更大的公差值，一般为 ±0.013 mm。

许多材料具有内应力，当这些材料用激光加工时，加工工件在切削过程中将会通过微小移动以消除内应力。在公差值允许范

stipulating tolerances.

Factors such as the acceleration and deceleration of the laser machine, part thicknesses, and the lagging effects of the laser cut, a slowing down and dwell time for sharp corners, is put into the program. Various slowing down and dwell times are used depending on material type and thicknesses. By allowing greater tolerances, the laser is permitted to cut faster.

5.1.5 Surface Condition

The surface condition of the material is important. Rust or scale on the surface impedes the cutting process and can cause irregular surfaces and excavations. Buying material such as pickled and oiled mild steel or cleaning the surface by sandblasting helps in solving the problem.

5.1.6 Beam Quality

Beam quality is one of most important factors for efficient laser operations. In laser cutting, the radiation power, radial distribution, and the circular pattern of the beam are important factors in maximizing the accuracy and effectiveness of the laser.

One method used for checking the laser beam is to allow the unfocused beam to melt a clear bar of acrylic plastic to a certain depth. To determine the laser beam cutting

围内，应考虑这一因素。

在加工时需要考虑的因素很多：激光切割机的加速或减速、工件厚度、激光加工的滞后现象、加工拐角时的减速或停留时间等。滞留时间是由材料厚度和类型决定的。当允许较大公差值时，激光加工的速度可以稍快一些。

5.1.5 表面质量

材料的表面质量是很重要的。加工过程中表面的铁锈和划痕会导致不规则表面或表面凹凸不平。我们可以通过使用盐润或油润的钢材或用喷砂器清洁其表面来解决这一问题。

5.1.6 激光束质量

激光束的质量是一个影响激光加工效率的重要指标。在激光加工中，射线的功率、射线的分配、光束的循环类型都是使激光加工的精确度和效率最大化的重要因素。

一项检测激光束的方法是使用激光束使乙烯纤维塑料熔化一个适宜的厚度。为了研究光束切割效率的决定性因素，我们

effectiveness, the burned shape pattern is analyzed.

Circular polarization of the laser beam has a strong effect on the cutting accuracy. To check for beam quality, eight slits should be cut in a piece of mild steel, each 45 degrees apart from the center. The bottom and top of the slits should be checked for any deviation. Cleaning, adjusting, or replacing a defective unit may be required to eliminate the deviations.

分析了激光束熔化塑料的形状类型。

激光束的循环极化现象会极大地影响切割精确度。为了检测激光束的质量，我们在一块钢材上面切出 8 个狭缝，每一个都与中心成 45° 角。检测狭缝的顶端和底端与理想位置的偏差，我们通常采用清洁表面、调整或更换有瑕疵的部件等措施来减小偏差。

5.2　Laser Component　激光加工器的组成

5.2.1　Resonator

A resonator is the unit which creates the high energy light beam which passes through optics. In an axial flow CO_2 laser resonator, voltage is applied to an anode and a cathode in a tube filled with helium, nitrogen, and CO_2 gases. In a 1500 watt laser, up to 20,000 volts of DC current excites the gases-raising the energy level of the lasing medium. When the lasing medium returns to its unexcited state, a photon of energy is given off. This photon collides with another excited atom of the lasing medium and produces another photon. In millionths of a second, a laser beam is created as the photons bounce back

5.2.1　共鸣器

共鸣器是产生通过光学镜片的高能激光束的部件。在 CO_2 激光器的共鸣器中，加电压于充满氦气、氮气、CO_2 气体的软管两端。在 1500 W 的激光切割器中，高达 20,000V 的直流电压加在气体上，这就提高了激光媒介所能产生的能量水平。当气体媒介回到非激发的阶段时，光子的能量就消失了。光子会与另一个被激发的原子在介质中相互碰撞并将产生另一个光子。在百万分之一秒内，随着光子在共鸣器和反光镜之间反复碰撞，激光束便

and forth between the resonator mirrors.

This stimulated emission of radiation: creates a chain reaction resulting in laser light. Helium inside the resonator acts as a cooling gas. The laser gases are circulated by means of a pump and cooled by a chiller.

The invisible laser light is monochromatic; it consists of only one color and one wavelength. The wave length travels in only one direction and coherently; i.e: the wave lengths are in phase with each other. Laser light photons travel an extremely narrow path. For example, a laser beam was aimed at a mirror on the moon. After making a round trip of half a million miles, the reflected laser light was picked up on earth.

5.2.2 Laser Mirrors

With resonators in the shape of a "U", the laser light inside the resonator bounces back and forth against reflective mirrors until it increases to an intensity that can pass through a partially reflective optic (PR mirror). These PR mirrors are typically from 35 to 80% reflective with the remaining transmissive. Light which has not passed through the partially reflective mirror bounces back, hitting the nitrogen gas which collides with the CO_2 gas. This collision then creates more photons to hit the mirrors. See Figure 5-4.

随之产生。

这种射线的激发辐射产生了导致激光产生的链式效应。在共鸣器中的氦气起冷却气体的作用。激光器中的气体是通过一个气体泵循环并被冷却器冷却。

这种不可见的激光束是单色的，它由单一颜色、单一波长的光组成。这种光波只有一个方向并且走向极其有条理，而且这些光波都是同相位的，激光光子的轨迹几乎是一条直线。例如，一束激光束被发射到月球上的一面镜子上。在往返 50×10^4m 之后，反射回的光线被地球所接收。

5.2.2 激光器的反光镜

共鸣器被设计成 U 形的同时，共鸣器内的激光束在反光镜间来回折射直至其能量足够通过一个部分反光镜（PR mirror）。这种部分反光镜可以反射剩余光线的 35%~80%。不能通过的光线将反射回来，撞击氮气和二氧化碳的混合气体。这种撞击现象将产生更多的光子来撞击反光镜，如图 5-4 所示。

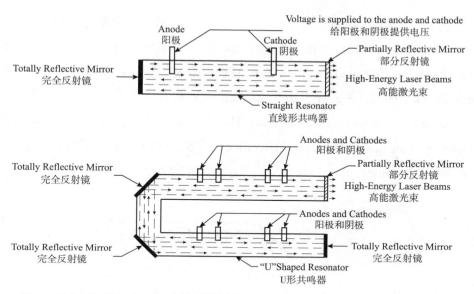

Figure 5-4　Straight resonator and resonator shaped as a "U" to produce laser beams
直线形共鸣器与 U 形共鸣器产生激光束的过程对比

The high-energy beams of light eventually pass through the partially reflective mirror and are directed by totally reflective mirrors (TR mirrors) to the laser nozzle as shown in Figure 5-5.

高能激光束最终通过部分反光镜并被全反射镜聚焦于喷嘴处，如图 5-5 所示。

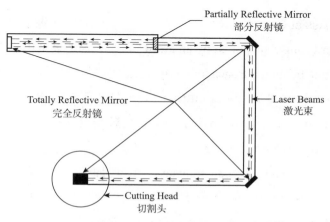

Figure 5-5　The transfer of laser light to the cutting head for a flying optic laser
激光束向切割起始点的传递

5.2.3 Laser Optics

A lens at the cutting nozzle focuses the laser beam as illustrated in Figure 5-6. The lens focuses the laser energy onto a small spot, which greatly increases its intensity. The thickness of the material being cut determines the size of the lenses.

5.2.3 激光器的光学镜片

喷嘴处的光学镜片采用如图 5-6 中所示方法聚焦了激光束。该镜片将光束的能量聚焦到一点上，这样就能极大地加强其强度。被切割材料的厚度决定了镜片的大小。

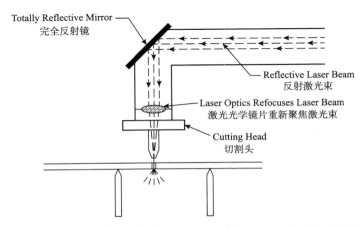

Figure 5-6　Laser optics focus the laser beam which creates intense energy of light
激光器光学镜片聚焦激光束，产生高能光束

5.2.4 Assist Gases

The laser beam is supported by an assist gas of air, argon, helium, nitrogen, or oxygen. When oxygen is used in metal cutting, it allows the laser to cut faster by assisting the laser beam to vaporize the molten material and to force out the molten material. Sometimes a stream of water or a water shield is used to prevent thermal effects when cutting with oxygen.

5.2.4 辅助气体

激光束是由空气、氩气、氦气或氧气等辅助气体支持。当氧气用于材料切割时，它通过辅助激光束蒸发已熔化的金属并熔化未熔化的金属来使激光器切削速度加快。有时候也使用水来阻止用氧气切割时材料产生的热膨胀效应。

When inert gases such as argon or nitrogen are used, they aid only to drive out the molten material from the gap; they do not aid the cutting process. Nitrogen is used to cut stainless steel to provide an oxide-free edge, sometimes called "clean cut". The advantage of "clean cut", though slower than oxygen cutting, is it eliminates a cleaning process to remove oxide scale from the cut surface. This is particularly important when stainless steel parts need to be welded or are to be used in a sanitary environment. Inert gases are also used to cut plastics, wood and other nonmetallic materials. See Figure 5-7.

当使用类似氩气或氮气的惰性气体时，它们仅仅起到清除加工间隙中的熔化材料碎屑的作用，对加工过程本身并没有什么帮助。氮气在切削不锈钢时能起到提供无氧区域的作用，这种加工也被称作"洁净加工"。尽管比有氧加工速率更慢，"洁净加工"还是有着自己的优势，其消除了从切割表面清洗除去氧化层的步骤。当不锈钢件要用于焊接或用于洁净的环境时，这一特点就起着很大作用。惰性气体也用于切割塑料、木材或其他非金属材料，如图 5-7 所示。

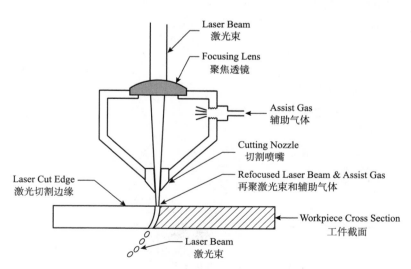

Figure 5-7　The laser beam and assist gas
激光束和辅助气体

5.2.5 The Laser Cut

High pressure assist gas quickly forces the molten material from the kerf. The molten material absorbs much of the heat; however, some heat is absorbed by the material, leaving a small amount of recast. This minimal resolidification of molten material is an important benefit of laser cutting, resulting in a minimal heat-affected zone and heat distortion.

The laser beam melts the material and helps to vaporize it as the assist gas forces out the molten mass. The combination of assist gas pressure and evaporation removes the molten material as shown in Figure 5-8. An exhaust fan and filter system removes the gases and fumes.

5.2.5 激光切割

高压辅助气体从切缝中快速喷过被熔化材料，被熔化的材料吸收了绝大多数热量；然而，一部分热量被材料本身吸收，这就导致了少量的材料出现重铸现象。熔化材料重铸量较小是激光加工的一个很大的优势，可产生较小的热影响区和热变形量。

激光束能在辅助气体将加工碎屑吹出加工间隙时熔化并蒸发材料。辅助气体综合了压力和蒸发作用，移除熔化材料的过程如图 5-8 所示。排气扇和过滤系统排除辅助气体和废气。

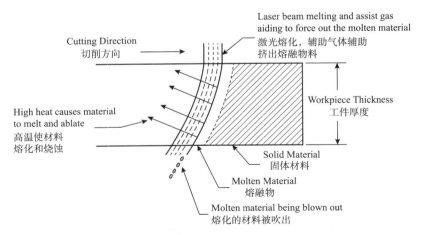

Figure 5-8　Laser beam melting the material as the asst gas force out the molten mass
激光束熔化材料，辅助气体移除被熔化碎屑

5.2.6 Sensing Unit

On some lasers, a height-sensing unit maintains the focal point at a fixed distance below or above the workpiece, even though the surface of the workpiece may fluctuate. This assures cutting uniformity. To prevent nozzle damage, some lasers are equipped with height scanners and a collision protection device. See Figure 5-9.

5.2.6 传感装置

在有些激光切割器中，即使工件的表面可能会波动，高度传感器还是能将焦点保持在一个距工件的上、下部某个固定的距离处，这就保证了加工的标准化。为了防止喷嘴损坏，有些激光切割器提供高度监测装置和冲击保护装置，如图 5-9 所示。

Figure 5-9　Laser nozzle equipped with a beam director for crash protection
　　　　　装有光束导向装置和冲击保护装置的激光切割器

5.3　Various Lasers and Their Configurations
　　　各种激光器及其配置

Two lasers dominate the material-processing field——the CO_2 laser and the Nd:YAG laser.

在切割材料的领域中有两种主要的激光切割器：二氧化碳激光切割器和钕钇铝石榴石激光切割器（YAG 激光切割器）。

5.3.1　How YAG lasers work

The Nd:YAG laser, (neodymium-

5.3.1　YAG 激光切割器的工作原理

YAG 激光切割器（钕钇铝

doped yttrium aluminum garnet) is a solid-state laser. Solid state means that the laser beam is produced by a crystal. The laser rod, which has been doped with an optical pure material, can lase. The yttrium aluminum garnet, a synthetically grown crystal, is doped with neodymium atoms. In contrast, the CO_2 laser is a gas laser. The CO_2 gas laser uses a mixture of helium and carbon dioxide within a chamber to obtain laser energy instead of a soild rod.

Flashlamps within the Nd:YAG lasers excite the media, in a process called pumping, which produces the laser beam. In Nd:YAG lasers, a very small amount of the electrical power sent to the flashlamps is converted to laser energy; most of the power turns to heat. The heat is withdrawn from the laser by a coolant system.

As laser beams emerge from the laser rod, the beams stricke the rear reflective mirror, then they reflect to the front of the partially reflective mirror. When beams of sufficient energy develop, they escape from the partially reflective mirror to form the laser beam used for production. See Figure 5-10.

Nd:YAG lasers come in continuous wave and pulsed mode. Flashing the lamp inside the laser produces the pulsing. Pulsed

石榴石激光切割器）是固基激光器，固基是指激光束是由水晶产生的。由纯光学玻璃制成的激光管可以发出激光。钇铝石榴石是一种掺有钕原子的复合式水晶。与之相反，二氧化碳激光器是一种气体基激光器。二氧化碳激光器没有使用固体管，而是使用二氧化碳和氮气的混合气体来储存能量。

YAG 激光器中的闪光灯：YAG 激光器在抽气过程中激发产生激光的媒介，从而产生激光束。在 YAG 激光器中，少量的能量被运送到闪光灯处转换成激光能量，一大部分能量转换为热量，这些热量被制冷系统排出激光器。

激光束从激光管中射出后会撞击后反射镜，然后反射到部分反射镜的前方。当激光束的能量足够时，他们将从部分反射镜中射出并用于加工，如图 5-10 所示。

YAG 激光器有连续模式和脉冲模式。激光器中的闪光灯能产生脉冲。脉冲模式可以使激光的

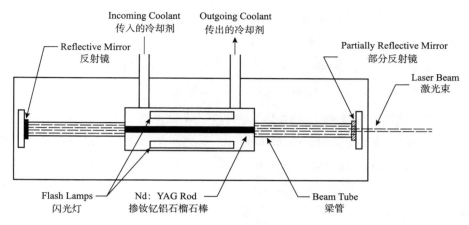

Figure 5-10 Nd:YAG laser
YAG 激光切割器

mode can increase laser peak power up to 500 times, and is particularly useful for drilling and welding.

5.3.2 Increasing Power for Nd:YAG Lasers

To increase the power for Nd:YAG lasers, individual Nd:YAG lasers are connected in series. Four 500-watt-lasers connected in series produce approximately 2000 watts of power.

For greater cutting power, the CO_2 lasers increase their power to 3000 watts. The advantage of this higher power is faster processing rates for thicker materials. Also, the wave length of the CO_2 laser aids in cutting non-metals.

最高能量提升到原来的 500 倍，通常用于打孔和焊接。

5.3.2 增加 YAG 激光器的功率

为了增加 YAG 激光器的功率，我们将许多台 YAG 激光器连在一起。四台 500 W 的激光器连在一起就可以提供 2000 W 的功率。

为了得到更大的功率，二氧化碳激光器将能量增加至 3000 W。高功率的好处是加快厚工件的加工速度。并且二氧化碳激光器的激光波长可以用于切割非金属。

5.3.3 Benefits of Nd:YAG Lasers

(1) Fiber Optics

One of the great benefits of Nd:YAG lasers is the ability of the laser beam to be used with fiber optics as shown in Figure 5-11. Some Nd:YAG lasers are capable of using a fiber optic cable that is over 600 feet (183m) long. This allows the user to reach into remote and difficult locations.

5.3.3 YAG 激光器的优点

（1）光纤

YAG 激光器的一大优点就是激光束可以使用光纤管传输，如图 5-11 所示。有些 YAG 激光器能够装载超过 600 英尺（约 183 m）的激光光纤软管，使用户能在更偏远或更困难的场合应用。

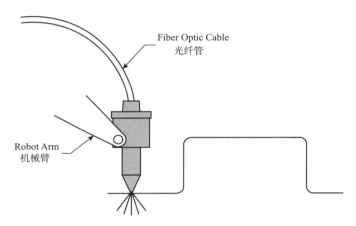

Figure 5-11　Nd:YAG laser with fiber optics trimming sheet metal
装载有光纤管的 YAG 激光器给钣金修边

This ability of lasers to use high-power-density beams through fiber optics allows for on-line production with robots. This fiber delivery system also allows multistation use by sharing the beam. This "time-sharing" of the laser beam is commonly used to minimize distortion during the welding process. Time-sharing also allows the laser to be used while loading and unloading.

激光加工将通过光纤使用高能激光束的能力运用到机器人的在线生产中。光线传送系统也允许通信站使用激光束传递信息。这种激光束的"共享"通常被广泛应用于焊接过程中以减少失真现象；这种共享也使激光束用于光盘的读取和刻录中。

With an articulated-arm or a five-axis robotic system, Nd:YAG can use a conventional positioning system with its fiber optics. However, CO_2 lasers generally deliver their beams to the workstation by means of an expensive specially engineered mirror system.

(2) Smaller Focusing Beam

Since a Nd:YAG laser beam is about one tenth the wave length of CO_2 lasers, it can focus to a very small diameter. This produces a higher energy concentration, making it ideal for drilling.

5.3.4　Excimer Lasers

Excimer lasers were first developed in 1975. The niche for excimer lasers is micromachining of ceramics, glass, polymers, plastics, metals, and diamonds. These lasers have been used for biomedical flow orifices, hybrid microelectronics, optoelectronics, diamond substrate machining, micromotors, fine wire stripping, and small optical apertures. See Figure 5-12.

5.3.5　How does Excimer Lasers Work

Basically, excimer lasers do not use a thermal process. They ablate material by breaking down the molecular bonds of the material, which produces a plasma plume.

装载有铰接式手臂或五轴联动机器人系统，YAG激光切割器可以用于接有光纤的传统定位系统中。然而，二氧化碳激光器通常使用非常昂贵的特殊反射系统来将其激光束传递至工作站。

（2）更小的光束聚焦

由于YAG激光切割器的波长是二氧化碳激光切割器二十分之一，所以它的光束可以聚焦到一个非常小的圆内。这样就产生了更高能量密度的光束，这样的光束是用于打孔的最理想光束。

5.3.4　准分子激光切割器

准分子激光切割器是在1975年发明的，其目的是方便陶瓷、玻璃、聚合物、塑料、金属和钻石的微细加工。这种激光切割技术已经被用于流孔生物医学、混合微电子、光电、金刚石基体加工、微型电动机、细线剥皮以及小型光孔加工，如图5-12所示。

5.3.5　准分子激光加工原理

首先，准分子激光加工不用热加工的方式，它是通过打破分子间的化学键连接而移除金属的，这种方式会产生一条等离

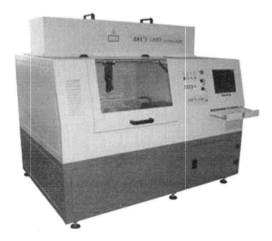

Figure 5-12　A micromachining excimer laser center
准分子激光微加工中心

The excimer laser, with its short wavelength, is essentially a heatless process, thereby producing virtually no heat-affected zones.

5.3.6　Capabilities of Excimer Lasers

By using a pulsed, deep ultra-violet excimer laser, hole diameters have been produced below 2 micron.

Perhaps the greatest benefits of excimer lasers is their capability in the production of high-density printed circuit boards. Hybrid circuits for personal computers, mainframes, and super computers require the removal of various polyimide insulating materials and the creation of microscopic circuit patterns.

子带。具有更短波长的准分子激光束进行切割的过程是一个非热加工过程，因此不会产生热影响区。

5.3.6　准分子激光加工的能力

通过使用深紫外线准分子激光脉冲，切出的孔洞直径可达到 $2\mu m$。

或许准分子激光加工最大的优点就是其加工高密度印刷电路板的能力。个人计算机、大型机及超级计算机的混合电路一般需要去除各种聚酰亚胺绝缘材料并创建微电路模式。

5.4 Traveling Methods of Laser Cutting Machines 激光切割器的移动方式

Lasers have three basic traveling methods: beam traveling, workpiece traveling, and combination of beam and workpiece traveling. Each system has its advantages and disadvantages.

5.4.1 Beam Traveling

The work table remains stationary while the laser beam travels. This process is called flying optics. The advantage of flying optics is the weight of the material being cut has no effect on machine movement. This allows for rapid travel when the machine is cutting, and when the machine is travelling to another section.

The disadvantage is that the laser beam can vary slightly in diameter from one corner of the machine to another. To test for possible variance, the same diameter hole should be cut in each of the four corners of the machine and the difference measured.

5.4.2 Workpiece Traveling

The laser remains stationary while the workpiece travels. The advantage of this method is that the laser beam size remains

激光切割器有三种移动方式：光束移动、工件移动、工件和激光束联合移动，每种方式都有其各自的优缺点。

5.4.1 光束移动方式

这种方式是加工平台保持静止而激光束本身移动，又称为移动光路型加工。此种方式的优势在于加工重工件时不必移动工件本身，这就实现了加工和切换加工面时的高频移动。

缺点在于激光束从机器一端移动到另一端时，光束的直径会发生轻微改变。为了测试改变量的值，需要在同一机器的四个角处用不同的方式切出同样直径的孔。

5.4.2 工件移动方式

这种加工方式是激光束不动而工件本身运动，优势在于激光束的粗细能保持恒定不变。这种

constant. This method is also used when lasers are combined with turret punch presses.

The disadvantage of the workpiece travelling method is that the weight of the workpiece can affect cut accuracy. To produce accurate cuts, slower table speeds are used.

5.4.3 Combination of Beam and Workpiece Traveling

A hybrid system is commonly used where the laser moves in the X direction and the table moves in the Y direction. The advantage is that there is a smaller variation in the beam path length. This stability produces greater accuracy than the beam traveling method.

The disadvantage is that it is not as fast in high-speed machining as the beam traveling method.

加工方式也用于激光束发生器和高压冲床连接使用时。

工件移动方式的缺点在于如果工件质量过大，加工精确度将难以保证。为了保证加工质量，常常降低工作台的移动速度。

5.4.3 混合移动方式

混合移动方式是激光、发生器沿 X 方向移动而工作台沿 Y 方向移动。优势在于激光本身的步长不会有太大的改变，这种稳定性与光束移动方式相比具有更高的精确度。

缺点是其加工速度比光束移动方式要慢。

5.5 Lasers Machining Applications 激光加工的应用

5.5.1 Laser Cutting

An understanding of the capabilities and advantages of lasers helps manufacturers utilize their many benefits. Listed below are some of the materials that can be cut with lasers (See Table 5-1).

5.5.1 激光切削

了解了激光加工的优势和能力，许多生产厂商能够最大限度地利用激光加工技术。

表 5-1 中列出了可以用激光加工的多种材料。

Table 5-1　Materials for laser cutting　能用激光切削的材料

Superalloys 超合金	Titanium 钛	Plastic 塑料
Stainless steel 不锈钢	Aluminum 铝	Kevlar 凯夫拉纤维材料
Tool steels 工具钢	Brass 黄铜	Wood 木材
Carbon steels 碳素钢	Copper 铜	Bakelite 电木
Galvanized steel 镀锌钢	Monel 蒙乃尔合金（镍合金）	Fiber 纤维
Hardened steel 硬质钢材	Beryllium copper 铍铜合金	Paper 纸
Inconel 铬镍铁合金	Ceramics 陶瓷	Masonite 绝缘纤维板
CPM 10V CPM 10V 材料	Screen materials 屏幕材料	Cork 软木
Hastelloy 哈氏合金	Composites 复合材料	Leather 皮革
Carbide 碳化物	Gasket materials 垫圈材料	Fused silica 熔融石英

Any material, whether metallic or nonmetallic, can be cut with lasers. There are two considerations in cutting various materials-the optical and the thermal characteristics of the material. The optical is the ability of the material to reflect the laser wavelength, whereas the thermal is the ability of the material to absorb the laser wavelength.

Since lasers cut by means of high energy light, some difficulties are encountered with highly reflective materials such as gold

任何的材料，不管是金属材料还是非金属材料，都可以用激光加工的方式加工。在切割诸多材料时，有两点因素需要考虑：材料的光学特性和热力学特性。光学特性是指材料反射光波的能力，而热力学特性是材料吸收光波的能力。

由于激光加工是通过高能激光束进行的，在加工金和银这些高反射性材料的时候会出现一些

and silver. For these materials, mechanical cutting methods are generally more cost effective. Also, the high thermal conductivity of aluminum and copper causes laser beam heat to be dissipated quickly. As a result, a 1500 watt CO_2 laser cuts only up to 3.2mm aluminum and up to 1.6mm copper. Laser wattage determines how thick a laser can cut. A 1500 CO_2 laser can cut carbon steel up to 13mm. A 3000 CO_2 watt laser can cut carbon steel up to 19mm.

Some lasers are equipped to cut tubes. The "Z" axis is programmable with a rotational axis. These three-axis lasers can cut various types of tubing, as pictured in Figure 5-13 and Figure 5-14.

问题。对这些材料来说，机械切削的方式更加经济适用。同样，由于铝和铜的高导热性，激光束会很快消散，所以1500 W的激光器只能切割3.2 mm厚的铝板或1.6 mm厚的铜板。激光功率大小决定了激光器可加工的材料厚度。1500 W的二氧化碳激光器可以切割厚度为13 cm的工件；3000 W的激光器能切割19 mm厚的碳素钢。

有些激光器用于切削管材。"Z"轴通常作为旋转轴，三轴联动激光切割器可以切削各种各样的管材，如图5-13和图5-14所示。

Figure 5-13　Laser cutting tubes
激光加工切削管材

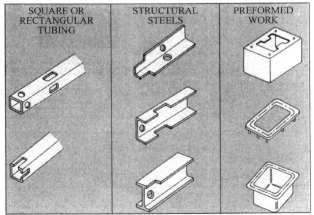

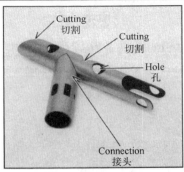

Figure 5-14　Various tube and preformed cutting applications
各种软管和预制工件的加工应用

5.5.2　Laser Welding

The two common lasers used for welding are Nd:YAG and CO_2. Welding occurs when a high intensity focused laser beam is directed and joins two or more similar or dissimilar metals by melting them together. See Figure 5-15. An inert gas of argon or helium generally surrounds the laser beam to protect the weld from oxidation.

Lasers are capable of performing deep welds known as keyhole welds. The welds

5.5.2　激光焊接

焊接时采用两种激光器：YAG 激光切割器和 CO_2 激光器。激光焊接其实就是指将两种或多种材料熔化成一体的过程，如图 5-15 所示。通常使用氩或氦等惰性气体充斥激光束的周围以保证加工部位不会被氧化。

激光加工能够加工出钥匙孔似的深焊缝，焊缝宽度也可以

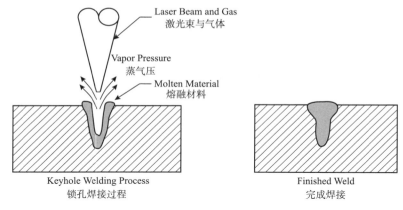

Figure 5-15　laser welding
激光焊接

can have a depth to width ratio of over 10:1. Keyhole welding occurs when a high power laser beam penetrates the metal and causes it to melt. The center of the weld begins to boil, while the outer edges of the molten material cling to the solid material. The surrounding cooler edges where the weld is being done have a higher surface tension; therefore the cooler edges pull the molten material to form a narrow pocket. Vapor pressure from the boiling material prevents the narrow pocket from collapsing.

5.5.3　Laser Cladding

Laser cladding adds material to a surface so it can withstand greater wear. The advantage of cladding is that an inexpensive material can be coated with an expensive alloy, such as stellite. In addition, the soft undercoat of the hardface makes the part

做至 10∶1。锁孔焊接发生在高功率激光光束穿透金属并且使它熔化之时。焊接区的中心开始沸腾，与此同时，熔化材料的外边缘开始凝固。焊缝边缘冷却下来之后会有更高的表面密度，因此冷却下来的边缘牵引被熔化的材料便形成一个窄孔。沸腾材料的蒸发压阻止了窄孔坍塌闭合。

5.5.3　激光镀

激光镀是往工件表面上镀一层材料，从而使工件更加耐磨。其优势在于可以将昂贵的材料（例如钨）镀在廉价的材料表面，从而减少成本。另外，硬镀层下的软材料使得工件的承载能

more durable.

Laser cladding is performed when a laser beam is focused on a desired surface which has been prepared with a powder substrate, or while a powdered alloy for hardfacing is applied, as depicted in Figure 5-16. The laser melts the applied alloy, and the alloy adheres to the underlying surface. Four different types of material are generally used for cladding: iron base, nickel base, cobalt base, and carbide.

力加强。

将镀层材料粉末撒在工件的表面，然后将激光聚焦在粉末上，将镀层材料熔化，冷却下来之后便完成了激光镀的过程，如图 5-16 所示。激光熔化镀层金属，然后镀层金属黏附到受镀表面上。我们通常使用四种材料作为镀层材料：铁、镍、钴、碳化物。

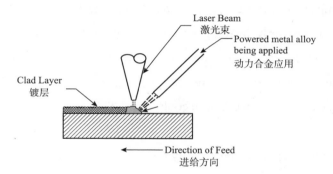

Figure 5-16 Laser cladding
激光镀

A process similar to laser cladding is laser surface melting. A thin layer of material is placed on the workpiece for the laser to melt. As the surface material melts and is "self-quenched", it creates beneficial surface properties on the workpiece.

5.5.4 Laser Heat Treating

Laser heat treating, or what is more precisely described as surface transformation hardening, is widely used to improve the

一种和激光镀类似的过程是激光表面熔化。我们将一薄层金属放置在工件表面，用激光熔化，随着表面材料熔化并"自我淬火"，工件的表面质量会得到显著提升。

5.5.4 激光热处理

激光热处理或者称为表面热硬化，在提高材料表面质量方面被广泛应用。

surfaces of machined parts.

Generally, laser beams emit an intense, narrowly focused beam of light. For heat treating, however, the laser beam is defocused in order to cover a larger area. The shape and size of the laser beam is achieved by using special optical systems designed for this purpose.

In laser heat treating, the laser beam heats the workpiece surface to above the austenitic temperature, but below melting. Since the surface is only heated, the cool interior of the workpiece quickly dissipates the heat when the beam passes. Quenching and hardening occurs as the workpiece cools.

Laser heat treating is suited primarily for localized hardening and on self-quenching metals. The depth of the case hardening and its hardness are determined by the workpiece material, the amount of applied laser energy, and the processing speed. See Figure 5-17.

5.5.5 Laser Marking

Lasers are used to mark hardened bearing cages, rulers, carbide drill bits, circuit boards, surgical instruments, labels requiring serializing, and many other products.

Lasers can engrave all sorts of metals, plastics, ceramics, and other materials on

通常情况下，激光加工的光束是一束高密度、小直径的光束。然而，对热处理而言，我们要使用激光束加工处理一大片区域。激光束的形状和宽窄通常是通过为此目的而特殊设计的光学镜片实现的。

在激光热处理过程中，激光束将工件的温度加热到奥氏体形成温度以上、熔化温度以下。由于我们只是加热表面，工件内部的未受热部分将在激光束经过之后将热量快速地传导出去。在工件冷却时，表面将发生淬火和硬化。

激光热处理主要用在材料的小范围硬化和自淬火方面。热处理的深度和硬度是由工件材料、激光能量、激光束的移动速度共同决定的，如图 5-17 所示。

5.5.5 激光打标识

激光可以用于标识硬质承载笼、尺子、钻孔机的钻头、电路板、外科手术用具、需要序列化的标签和许多其他产品。

激光可以雕刻所有类型的金属、塑料、陶瓷和其他材料，并

Unit 5　Lasers Machining　激光加工 | 107

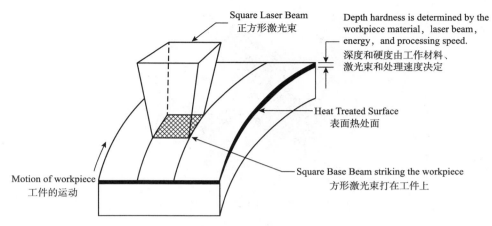

Figure 5-17　Laser heat treating
激光热处理

both flat and round surfaces. See Figure 5-18.

5.5.6　Laser Drilling

Laser drilling performs close-tolerance high-speed drilling in practically all materials, from super alloys to soft rubber. Some lasers can drill up to 50 holes a second. See Figure 5-19 for a laser drilled gas turbine.

且无论其表面是平坦的还是圆形的都可以加工，如图5-18所示。

5.5.6　激光钻孔

激光钻孔可以实现从超合金到合成橡胶材料的小公差高速钻孔加工。有些激光器可以在1秒钟内钻出50个孔。图5-19所示的是一台用激光钻孔的涡轮机。

Figure 5-18　Various laser-marked parts
种类繁多的激光标识产品

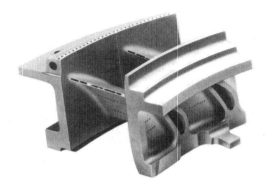

Figure 5-19　Laser drilled gas turbine
采用激光钻孔的涡轮机

Questions　习题

1. What is one of the major reasons lasers are so cost effective?

2. What does the word "laser" stand for?

3. Describe how a laser light is created in the resonator.

4. What is the purpose of laser optics?

5. Describe how Nd:YAG lasers work.

6. How long a cable can be used with fiber optics?

7. List the three basic traveling methods for lasers and describe each.

8. List four applications of laser machining.

1. 激光加工能节省成本的主要原因之一是什么?

2. "激光"这个词代表了什么?

3. 描述激光是如何在共鸣器中产生的。

4. 激光器中光学镜片的作用是什么?

5. 描述 YAG 激光切割器的工作过程。

6. 带有光学纤维的电缆的长度最长可以是多少?

7. 列举激光加工的三种移动方式,并分别描述。

8. 列举激光加工的四种应用。

Photochemical Machining
光化学加工   6

6.1 Fundamentals of Photochemical Machining
光化学加工的基础

Photochemical Machining (PCM), also known as photochemical milling, photo-etching and photolithography, is a process that blanks or etches out parts by means of chemicals, as opposed to using abrasives or hard tooling. This technique avoids burrs, no mechanical stresses are built into the parts and the properties of the metal worked are not affected. Hardened and tempered metals are machined as easily as regular metals. The technique is ideal for machining thin metals and foils. Parts with very precise and intricate designs can be produced without difficulty. The photochemical machining/milling processes can precisely etch lines and spaces on all types of metals (kovar, nickel, brass,

光化学加工即所谓的光化学研磨、光刻法和影印石板术，是一种通过化学手段擦除或蚀刻工件的方法，而非传统上利用硬质工具的加工过程。这种技术能避免留下毛边，不在工件中产生化学应力并且不受加工金属材料的影响，高硬度金属也可以像普通金属一样加工。这种技术是加工薄细金属工件和箔的理想手段。要求精度很高且设计复杂的工件都可以利用这种技术轻而易举地加工。光化学加工可以在所有类型的金属（包括镍钴合金、镍、黄铜、铍、铜、不锈钢和铝等）上精确地划线定位。这种技术配

beryllium, copper, stainless steel, aluminum, and others) with detailed accuracies. This is used for creating specialty flex circuits, plus in engineering of other rigid technologies. This results in a burr free part with very close tolerances.

6.1.1 Designing the Part

Normally, a computer-aided design (CAD) is made for the desired share, and an etch line is put around the shape in order to remove the part from a metal sheet. The usual width of this line is twice the metal thickness. Often small tabs are used to hold the part during processing. After the design is completed, it is then plotted.

6.1.2 Imaging

The actual size of the image is then projected with a precision process camera onto film or a glass master. Glass masters are used for precision work because they are dimensionally stable and flat. Multiple negatives of the same image can be made with a process camera, or they can be reproduced on a photo repeating machine from the film or glass master. To create high-precision images, a process using laser-controlled photo plotters is used.

The metal surface to be etched is first cleansed to receive a photosensitive resist

合其他精加工手段可以用来加工特殊的柔性电路。

6.1.1 工件设计

通常利用计算机辅助设计来获得需要的形状，沿着工作的形状划一条蚀刻线从金属片上切下工件。这条线的宽度通常是金属厚度的两倍，并常常在设计过程中利用一些小标签来标号，在设计完成后，用它们来标定位置。

6.1.2 图像

图像的实际尺寸是利用一个精确制版照相机来照在胶卷上或一个玻璃基板上。玻璃基板由于其尺寸稳定且光滑平坦常被用来做一些精度工作。同一个图像的阴性图像可以利用制版相机来获得，或者从胶卷或玻璃基板上利用照相复印机来重塑。为了获得高精度图像，常利用激光控制绘图仪来做这个工作。

要被蚀刻的金属表面首先要清理并涂一层抗光敏涂层。图

Unit 6　Photochemical Machining　光化学加工 | 111

coating. Figure 6-1 shows coil coating of liquid resist for high volume production.

The negative images from the film or glass masters are then exposed on the photo resist-coated metal and developed. The image may be produced on either one side or both sides of the coated metal. See Figure 6-2 for a developer system.

6-1 是用来批量生产的液体光阻卷材涂料。

胶卷和玻璃基板上的阴性图像紧接着会在光阻金属涂层上曝光并做进一步处理。这样图像可能会在涂层金属的一面或两面都成像，图 6-2 是一个显像系统。

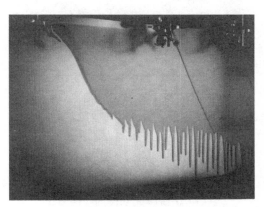

Figure 6-1　Coil coating of liquid photo resist for high volume production
用来批量生产的液体光阻卷材涂料

Figure 6-2　A developer system
显像系统

6.1.3　Etching and Stripping

The metal is chemically processed using a corrosive solution that is sprayed under

6.1.3　蚀刻和褪层

被加工金属要用一种腐蚀溶液来进行化学处理，即将该溶液

pressure onto the photo resist coated metal. The constant force of the spray washes away the reaction products. Where the photo resist has been exposed, the metal is etched away. See illustrations Figure 6-3 and Figure 6-4.

高压喷在光阻涂层金属表面。喷雾的持续力将反应产物冲刷掉。光阻被曝光的地方金属发生蚀刻,如图6-3和图6-4所示。

Figure 6-3　Spray system washing away exposed areas
用来清洗曝光部位的喷涂系统

Figure 6-4　An etching system
蚀刻系统

Depending on the duration of the chemical process, the etching can cut entirely through the metal or just scribe lines on

依靠化学反应的持续进行,蚀刻可以完全切断金属或只在金属表面划线。在工件被化学蚀刻

the metal surfaces. After parts have been chemically etched, they undergo a stripping process where the photo resist is stripped off the metal surface. See Figure 6-5 for the photochemical machining process.

后，在有光阻涂层的地方就发生了褪层过程，图 6-5 是光化学加工过程。

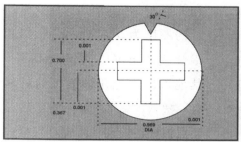
(a)Part is drawn with CAD
利用计算机辅助设计工件

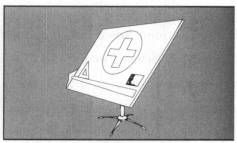
(b)Enlarged master drawing is secured with tabs
放大基板用来保护标签

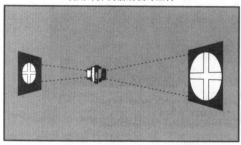
(c)Image is reduced to exact size and exposed on film
图像获得精确曝光

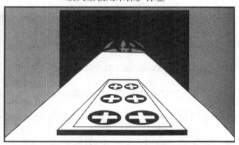
(d)Multiple images are exposed to a metal coated with photo resist
成批图像在光阻金属表面曝光

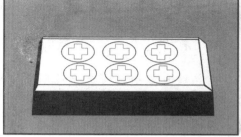
(e)A chemically corrosive solution is sprayed on the metal
化学腐蚀溶液喷涂在金属表面

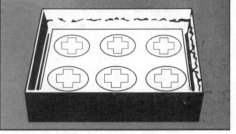
(f)Part goes through a stripping Process to remove chemicals All unprotected areas will be etched away.
工件经历褪层过程，所有没有被保护的地方将被蚀刻掉

Figure 6-5　The photochemical machining process
光化学加工过程

6.1.4 Materials and Products for Photochemical Machining

Many metals can be chemically etched, such as: brass, copper, nickel, silver, stainless steel, carbon steel, bronze, aluminum, tungsten, molybdenum, titanium, and zirconium. Some metals have been chemically machined from 0.13 to 1.57mm. Copper and aluminum alloys can be machined up to 2.54mm. Other materials that have been photochemically machined are: plastics, polyesters, polyamides, epoxy resins, glass, and ceramics. Various etchant chemicals and photoresists are used to make this process possible on such materials.

The printed circuit industry is a major user of this process. Other industries use this process in producing parts, such as: motor laminates, encoder discs, flapper valves, shims, contacts, screens, springs, heat sinks, decorative plaques, and ornaments.

6.1.5 Tolerances

A number of variables can affect the tolerances of parts: sheet and part size, type of metal and its thickness, and processing method. For general purposes, a tolerance of ±20% of the metal thickness can be used. Tighter tolerances can be obtained for certain designs. The generally stated relationship

6.1.4 光化学加工的材料和产物

很多材料可以被化学蚀刻，例如：黄铜、镍、银、不锈钢、碳钢、青铜、铝、钨、钼、钛、和锌等。一些金属可采用化学加工的厚度为0.13~1.57mm，钼和铝合金甚至可以达到2.54mm。其他可以用光化学加工的材料有：塑料、聚酯纤维、聚酰胺、环氧树脂、玻璃和陶瓷等。很多化学蚀刻剂和光阻材料都可以用来加工这些材料。

印刷电路行业是这种技术的主要使用者。其他利用这种技术的行业主要有：加工马达层压板、编码盘、插板阀、薄垫片、接头、屏幕、弹簧、散热板、装饰牌匾和装饰物等。

6.1.5 公差

很多变量可能会影响到工件的公差：薄层和工件尺寸、金属类型和它的厚度以及加工方式等。通常情况下，可以使用金属厚度的±20%作为公差。而更小的公差可以通过某种设计得到。通常情况下孔的直径和金属厚度

of the hole diameter to metal thickness is that the hole cannot be less than the metal thickness.

的关系是孔直径不能低于金属板厚度。

6.2 Advantages of Photochemical Machining 光化学加工的优点

(1) Eliminates the Need for Hard Tooling

Hard tooling is expensive, especially when there are many intricate shapes and holes.

(2) Just In Time Machining

Rapid turnarounds are possible since no hard tooling is required. The production can be accomplished in one day.

(3) Freedom of Burrs

Photochemical machining produces no burrs.

(4) Stress-Free Machining

When parts are blanked, stresses can be introduced into the material. Since there are no forces being applied with chemical etching, the metal remains flat. This is particularly important when machining thin pieces. Also, work hardening does not occur with this process.

(5) Delicate and Complex Parts Can be Produced

The complexity of the part does not

（1）不需要硬质工具

硬质工具非常昂贵，尤其是加工有复杂形状和孔的工具。

（2）及时加工能力

由于不需要硬质工具，所以可以进行快速循环使用。加工开始到得到产品可以在1天内完成。

（3）不产生毛边

光化学加工不产生毛边。

（4）加工时不产生应力

当利用机械加工等手段加工时，有可能在材料中产生应力。而在化学蚀刻过程中则没有应力，且金属板加工中仍能保持平整。这在加工薄片零件时非常重要，而且这种加工技术不会产生机械硬化现象。

（5）可以加工微小和复杂形状的工件

工件的复杂性并不影响加工

affect the cost or the processing time. Practically any drawn shape can be machined. With other processes, such as lasers and wire EDM, the length of cut is an important cost factor, but not with photochemical machining. The film is made with one exposure, and the etching process is done at one time.

Extremely complex and delicate parts can be produced with very narrow spaces because of the stress-free nature of the machining process.

所需要的成本和时间。利用这种加工技术几乎可以加工绘制的任何形状。利用其他加工手段，例如用激光和线切割切削的长度是影响成本的一个重要因素，而对于光化学加工则不是。薄膜一次曝光，蚀刻过程一次完成。

极其复杂和微小的工件利用光化学加工都可以在非常窄小的空间里完成，这是利用了加工过程中不存在应力这一优点来完成的。

6.3 Disadvantages of Photochemical Machining
光化学加工的缺点

(1) Bevel Slots and Holes

Photochemical machining always leaves a beveled edge, but sometimes parts require straight edges. For parts with multiple holes or slots, lasers are a good machining substitute. For thin parts without many holes, stacked sheets can be cut profitably with wire EDM, which also provides a perfect, burr free straight edge.

(2) High Run Production

When there is high production for some applications, it can become more cost effective to produce hard tooling.

（1）斜槽和斜孔

光化学加工常常会留下一个斜边，但有时一些工件需要直边。对于那些有很多孔和槽的工件，激光加工则是最好的选择。对于那些没有孔的薄件堆片可以利用线切割高效切削来获得不产生毛边的直边。

（2）大批生产

当在某些大批量生产作业中使用该技术时，它的生产成本甚至会超过硬质工具。

(3) Limited Metal Thicknesses

Although photochemical machining can cut copper and aluminum, up to 2.54mm, it is generally used to cut thin materials. For thick materials, other processes are used.

（3）被加工金属厚度有限

尽管光化学加工可以用来切削铜和铝，但也只能切割厚度不超过2.54mm的。它主要用来切割薄片金属，对于超厚的材料就只能用别的加工手段了。

6.4 Applications of Photochemical Machining 光化学加工的应用

A wide range of functional products can be made by photochemical machining in electronic, computer, metal working and precision engineering industries. Some typical applications worthy of meaning are as follows:

(1) Colour TV receiver tube aperture mask, commonly known as shadow mask, normally comprises of some 300,000 perfectly etched slots. Photochemical machining is the only feasible method for the shadow mask manufacturing.

(2) Integrated circuit lead fame, an electronics component with a complex geometry, is broadly fabricated by photochemical machining because of the rapid change in IC designs and also the geometry becomes more and more complex.

(3) Stainless steel spring and magnetic

光化学加工可以用来加工很多种类的功能性产品，如电子产品、计算机零件和一些精度要求高的产品等。一些典型的应用如下：

（1）彩色电视机接收器管的空隙罩子，即所谓的影孔罩板，通常被加工大概300,000个狭窄的蚀刻槽。光化学加工是在影孔罩板产业唯一可行的加工手段。

（2）集成电路的布置。它包含着大量的电子元件和复杂的几何学问题，也通常用光化学加工来完成。伴随着工业设计的快速发展和几何问题的愈加复杂性，光化学加工显得越来越重要。

（3）不锈钢弹簧和软驱磁头

head for floppy diskette drive are commonly fabricated by photochemical machining.

(4) Metal mesh screen, used in juice extractor, is another major photochemical machining application.

(5) Metal toy models, a small batch production of great number of different designs, are typical products made by photochemical machining.

(6) Decorative and graphic products are commonly made by photochemical machining due to the low volume and normally complex pattern design.

也通常用光化学加工来完成。

（4）榨汁机的金属筛网加工是光化学加工技术的重要使用场合。

（5）金属玩具模型的大批量生产，是光化学加工的典型应用。

（6）低产量但设计复杂的装潢和图案产品通常也用光化学加工来完成。

Questions　习题

1. What are some of the other names of photochemical machining?
2. After the material is coated, how is the image produced on the material?
3. Describe how the etching and stripping process creates the desired part.
4. List some of the applications of photochemical machining.
5. What is the advantages of photochemical machining?

1. 光化学加工的其他名称是什么？
2. 当材料涂层以后，怎样将图形制作在材料上？
3. 请描述一下如何蚀刻和褪层来获得想要的产品？
4. 请列举光化学加工的应用。
5. 光化学加工的优点是什么？

Ultrasonic Machining
超声波加工

7.1 Fundamentals of Ultrasonic Machining
超声波加工原理

Ultrasonic machining (USM) is the removal of hard and brittle materials using an axially oscillating tool at ultrasonic frequencies (20~50 kHz) with the abrasive. Because the process is nonchemical and nonthermal, materials are not altered either chemically or metallurgically. Ultrasonic machining is able to effectively machine all materials. Harder than HRC 40, whether or not the material is an electrical conductor or an insulator.

The ability of an ultrasonically vibrating tool to interact with an abrasive mixture and produce material removal was first discovered in 1950 by the American engineer Lewis Balamuth (Ensminger, 1973). Since

超声加工是利用轴向振动工具和磨料在超声频率（20~50 kHz）下去除硬脆材料的方法。因为这个过程是非化学性和非热性的，所以材料不会发生化学或冶金改变。超声波加工能有效地加工比HRC40更硬的材料，不管是电导体还是绝缘体。

1950年，美国工程师Lewis Balamuth（Ensminger,1973）首次发现了超声波振动工具与磨料混合物相互作用可产生材料去除的能力。自发明以来，超声波加工

its invention, USM has developed into a process that is relied upon to solve some of the manufacturing community's toughest problems.

The USM process begins with the conversion of low-frequency electrical energy to a high frequency electrical signal, which is then fed to a transducer. The transducer is a device that converts the high-frequency electrical signal to high frequency linear mechanical motion.

The high frequency mechanical motion is transmitted to the tool via a mechanical coupler known as the toolholder. The tool vibrates with a total excursion of only a few hundredths of a millimeter (thousandths of an inch) in a direction parallel to the axis of the tool feed. For efficient material removal to take place, the tool and toolholder must be designed with consideration given to mass and shape so that resonance can be achieved within the frequency range capability of the USM machine. Resonance is achieved when the frequency of vibration matches the natural frequency needed to generate a standing sonic wave within the tool-toolholder assembly, thus resulting in maximum vibrational amplitude and maximum material removal efficiency.

（USM）已经发展成为制造业所依赖用以解决一些最棘手问题的方法。

USM过程首先将低频电能转换为高频电信号，然后将其送入传感器。传感器是一种将高频电信号转换成高频线性机械运动的装置。

高频机械运动通过一个称为"刀柄"的机械耦合器传递给刀具。该工具在平行于刀具进给轴线的方向上仅以百分之几毫米（千分之一英寸）的总偏移振动。要实现有效的材料去除，刀具和刀柄的设计必须考虑到质量和形状，以便在USM机器的频率范围内实现共振。当振动频率与刀具–刀柄内产生驻声波所需的固有频率相匹配时，可实现共振，从而产生最大的振动幅度与最大的材料去除效率。

The tool is shaped conversely to the desired hole or cavity and positioned near, but not touching, the surface of the workpiece. The gap between the vibrating tool and workpiece is flooded with an abrasive slurry comprising water and small abrasive particles.

Material removal occurs when the abrasive particles, suspended in the slurry between the tool and workpiece, are struck by the downstroke of the vibrating tool. The impact from the tool propels the abrasive particles across the cutting gap causing them to strike the workpiece with a force up to 150,000 times their weight. Although this force sounds large, the small mass of the abrasive particles results in a very low overall cutting force, rarely exceeding 4.5 kg.

If the material being machined is brittle rather than ductile, a small crater will be produced at each impact site, much like the crater that results when a sheet of glass is struck by a BB or pellet from an air gun. Each downstroke of the tool can simultaneously accelerate thousands of abrasive particles; thus literally millions of chins are removed from the workpiece each second. This explains why the second most common name for this process is ultrasonic impact grinding (UIG).

该工具的形状与所需的孔或空腔相反，位置靠近，但不接触工件的表面。振动工具和工件之间的间隙充满含有水和小磨粒的磨料砂浆。

当悬浮在工具和工件之间的浆料中的磨料颗粒被振动工具的下行程撞击时，就会发生材料去除。来自工具的冲击推动磨粒穿过切割间隙，使得它们以高达其重量150,000倍的力撞击工件。尽管这种力听起来很大，但研磨颗粒的小质量导致整体切削力很低，很少超过4.5 kg。

如果被加工的材料是脆性的而不是韧性的，那么在每个冲击位置都会产生一个小的凹坑，就像一片玻璃被BB枪或气步枪击中的弹坑一样。该工具每次下冲程都可以同时加速数千个磨料颗粒，因此工件上每秒钟都会移除数百万个微粒。这解释了为什么这个工艺的第二个最常见的名称是超声波冲击磨削（UIG）。

A second, but much less effective mode of material removal takes place simultaneously with the first. A minor amount of workpiece material is abraded as a result of its interaction with the turbulent abrasive slurry.

As material is removed, a counterbalanced gravity feed, or servomotor-driven feed mechanism, continuously advances the tool into the newly formed hole to maintain a constant gap between the tool and workpiece. Although USM volumetric material removal rates are relatively low, the process remains economically competitive because of its ability, with a single pass of the tool, to generate complex cavities or multiple holes in workpiece materials that are too hard or fragile to machine by alternate processes. Additionally, because there is no direct tool-to-workpiece contact, USM is a valuable process for reducing manufacturing losses caused by in-process breakage of fragile workpieces.

In addition to its manufacturing capabilities, USM is also quite possibly the safest of all nontraditional and conventional processes because is involves no high-voltage, burning, cutting, chemicals, or dangerous mechanical motions. In fact, USM

第二种效率较低的材料去除方式与第一种同时进行。少量工件材料由于其与湍流磨料浆料的相互作用而磨损。

当材料被去除时，平衡的重力进给或伺服电机驱动的进给机构不断地将工具推进到新形成的孔中，以保持工具和工件之间的恒定间隙。尽管USM体积材料的去除率相对较低，但由于刀具能够单次通过，在材料太硬或太脆弱以至于不能通过交替工艺加工的工件中可加工出复杂的空腔或多个孔，所以该工艺在经济上仍然具有竞争力。此外，由于刀具与工件没有直接的接触，所以USM是一个有价值的工艺，可以减少因易碎工件在加工过程中损毁造成的制造损失。

除了制造能力之外，USM也可能是所有非传统和传统工艺中最安全的，因为它不涉及高压、燃烧、切割、化学或危险的机械运动。事实上，由于皮肤的延展性，USM甚至不能切割皮肤。

cannot even cut skin because of the skin's ductility.

7.2 Ultrasonic Machine 超声波加工机

The basic mechanical structure of an USM is very similar to a drill press. However, it has additional features to carry out USM of brittle work material. The workpiece is mounted on a vice, which can be located at the desired position under the tool using a 2 axis table. The table can further be lowered or raised to accommodate work of different thickness. The typical elements of an USM are See Figure 7-1.

USM 的基本机械结构与钻床非常相似。但是它还具有实现对脆性工作材料使用的附加功能。工件安装在虎钳上，它可以使用 2 轴工作台定位在刀具下的所需位置。工作台可以进一步降低或升高以适应不同厚度的工作。USM 典型元素如图 7-1 所示。

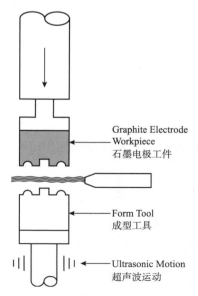

Fig 7-1 Major elements of ultrasonic machining
超声波加工的主要元素

7.2.1 Power Supply

The power supply used for USM is more accurately characterized as a high-power sine-wave generator that offers the user control over both the frequency and power of thus generated signal. It converts low-frequency (60 Hz) electrical power to high frequency (approximately 30 kHz) electrical power. This electrical signal is then supplied to the transducer for conversion into mechanical motion.

7.2.2 Transducer

A transducer is a device that converts energy from one form to another. In the case of transducers for USM, electrical energy is converted to mechanical motion. The two types of transducers used for ultrasonic machining are based on two different principles of operation, piezoelectric and magnetostrictive.

Piezoelectric transducers used for USM generate mechanical motion through the piezoelectric effect by which certain materials, such as quartz or lead zirconate titanate, will generate a small electric current when compressed. Conversely, when an electric current is applied to one of these materials, the material increases minutely in size. When the current is removed, the

7.2.1 电源

用于 USM 的电源可更精确地描述为一个高功率的正弦波发生器，它能为用户提供对所产生信号的频率与功率的控制。它将低频电力（60 Hz）转换为高频电力（约 30 kHz），然后将该电信号提供给传感器以转换成机械运动。

7.2.2 传感器

传感器是将能量从一种形式转换为另一种形式的装置。用于 USM 的传感器是将电能转换成机械运动。用于超声波加工的两种类型的传感器基于两种不同的操作原理，即压电和磁致伸缩。

用于 USM 的压电传感器通过压电效应产生机械运动，通过该效应，某些材料（例如石英或锆钛酸铅）在压缩时将产生小的电流。相反，当电流施加到这些材料中的一种时，材料尺寸会略增加；当电流消失后，材料立即恢复到原来的形状。压电式传感器本质上具有极高的机电

material instantly returns to its original shape. Piezoelectric transducers, by nature, exhibit an extremely high electromechanical conversion efficiency (up to 96%), which eliminates the need for water cooling of the transducer. These transducers are available with power capabilities up to 900W.

Magnetostrictive transducers are usually constructed from a laminated stack of nickel or nickel alloy sheets which, when influenced by a strong magnetic field, will change length. Magnetostrictive transducers are rugged but have electromechanical conversion efficiencies ranging from only 20 to 35%. The lower efficiency results in the need to water-cool magnetostrictive devices to remove the waste heat. Magnetostrictive transducers are available with power capabilities up to 2400 W.

The magnitude of the length change that can be achieved by both piezoelectric and magnetostrictive transducers is limited by the strength of the particular transducer material. In both types of transducers. the limit is approximate 0.025 mm.

7.2.3 Toolholders

The obvious function of the toolholder is, as the name implies, to attach and hold the tool to the transducer. Additionally, the

转换效率（高达96%），不需要水冷却。这些传感器的功率可达900 W。

磁致伸缩传感器通常由镍或镍合金片的叠层构成，当受到强磁场影响时，其会改变长度。磁致伸缩传感器坚固耐用，但其机电转换效率从20%到35%不等。较低的效率导致需要水冷却磁致伸缩装置以去除余热。磁致伸缩传感器的功率可达2400 W。

压电传感器和磁致伸缩传感器长度变化的大小受特定传感器材料强度的限制。在这两种传感器中，极限值约为0.025 mm。

7.2.3 刀架

顾名思义，刀架的显著功能是将刀具连接并固定在传感器上。另外，刀架还将声能传送到

toolholder also transmits the sonic energy to the tool, and in some applications, also amplifies the length of the stroke at the tool. In this capacity, the toolholder must be detachable. Be constructed of materials that have good acoustic properties, and be highly resistant to fatigue cracking.

Toolholders are attached to the transducer by means of a large, loose-fitting screw which has intentionally oversize threads in the female portion and undersize threads in the male portion. If this screw junction were tight fitting, an ultrasonic weld would permanently bond the toolholder to the transducer. Half hard copper washers are used between the transducer and toolholder to dampen and cushion the interface, thus further reducing the chances of unwanted ultrasonic welding. Regardless of the quality of the construction, all mechanical joints between the transducer, toolholder, and tool exhibit some amount of mechanical impedance resulting in a small amount of power loss and high local stresses. The presence of these high stresses leads to an increased probability of fatigue cracking.

The materials most often used to construct toolholders are Monel, titanium,and stainless steel. While titanium has the best acoustical properties, it cannot be successfully

刀具，并且在一些应用中还放大刀具的行程长度。在这种情况下，刀架必须是可拆卸的。其由具有良好声学性能的材料构成，并且具有高度抗疲劳开裂性。

刀架通过一个大的、宽松的螺钉连接到传感器上，该螺钉在凹入部分中有加粗的螺纹，在凸出部分中有过小的螺纹。如果这个螺纹接头很紧，超声波焊接将永久性地将刀具固定到传感器上。在传感器和刀架之间使用半硬铜垫圈来减震和缓冲接口，从而进一步减少不必要超声波焊接的可能性。无论结构和质量如何，传感器、刀架和刀具之间的所有机械接头都表现出一定量的机械阻抗，造成了少量的功率损失和较高的局部应力。这些高应力的存在使疲劳开裂的可能性增加。

最常用来制造刀架的材料是蒙乃尔合金（Monel）、钛和不锈钢。虽然钛具有最好的声学性能，但它不能被成功地钎焊，因

brazed and hence must rely on mechanical attachment of tools.

The acoustic properties of Monel are almost as good as titanium. In addition, tools can be brazed to Monel, thus making this the most often-used toolholder material.

Although it offers the benefit of being the least-expensive material choice, the acoustical properties of stainless steel are not as good as those of titanium or Monel, and because of its low fatigue strength, it is limited to low-amplitude applications.

Toolholders are available in two configurations: non-amplifying and amplifying. Non amplifying toolholders are cylindrical and result in- the same stroke amplitude at the output end as at the input end. Amplifying toolholders have a modified cross-section, and are designed to increase the amplitude of the tool stroke as much as 600%. Amplifying toolholders increase tool motion through stretching and relaxation of the toolholder material.

Because of the gain in tool stroke, amplifying toolholders are able to remove material up to 10 times faster than the non amplifying types. The disadvantages of using amplifying toolholders include increased cost to fabricate, a reduction in surface finish

此必须依靠刀具的机械连接。

蒙乃尔的声学特性几乎和钛一样好。此外，刀具可以钎焊到蒙乃尔上，从而使其成为最常用的刀架材料。

虽然不锈钢为最便宜的材料，但是它的声学特性不如钛或蒙乃尔合金，并且由于其低疲劳强度，只能限于低振幅的应用。

刀架有两种配置：非放大和放大。非放大的刀架是圆柱形的，并在输出端与输入端产生相同的行程振幅。放大刀架具有改进的横截面，是为将刀具行程的幅度增加至600%而设计的。放大刀架通过刀架材料的拉伸和松弛来增加刀具运动。

由于刀具行程的增加，放大刀架的材料移除速度能比非放大型快10倍。使用放大刀架的缺点包括制造成本增加，表面光洁度质量降低，以及需要频繁调整以维持共振。

quality, and the requirement of much more frequent tuning to maintain resonance.

7.2.4 Tools

To minimize tool wear, tools should be constructed from relatively ductile materials such as stainless steel, brass, and mild steel. Whenever possible, USM tools to be used for hole drilling are constructed from easily obtained materials such as music wire, stainless steel tubing, or hypodermic needles. Solid tools used to produce cavities can be fabricated by machining, casting, or coining; however finishing or polishing operations are sometimes necessary because the surface finish of the tool will be reproduced in the workpiece. Both tools and toolholders should be free of scratches, nicks, and heavy machining marks, because these produce stress risers and lead to early fatigue failure.

Because of the overcut that occurs with this process, allowances must be made to use tools that are slightly smaller than the desired hole or cavity. For example, to allow for overcut, the diameter of the tubing used to drill holes should be equal to the desired hole diameter minus twice the abrasive particle size.

The most desirable method of attaching the tool to the toolholder is by silver brazing.

7.2.4 刀具

为了减少刀具的磨损，刀具应该由不锈钢、黄铜和低碳钢等相对易延展的材料制成。在可能的情况下，用于钻孔的 USM 工具由易获得的材料构成，如琴弦、不锈钢管或皮下注射针。用于制造腔体的固体工具可以通过机加工、铸造或冲压来制造。然而，有时也需要精加工或抛光操作，因为工具的表面光洁度将在工件中再现。刀具和刀架都应该没有划痕、缺口和严重加工痕迹，因为这些会产生应力集中并导致早期疲劳失效。

由于在此过程中发生了过度切割，所以必须使用比预期的孔或腔略小的刀具。例如，为了允许过度切割，用于钻孔的管道直径应该等于所期望的孔直径减去研磨粒径的两倍。

将刀具连接到刀架上最理想的方法是银钎焊，这消除了机

This eliminates the fatigue problems associated with mechanical screw attachment methods.

7.2.5 Abrasives

Several abrasives are available in various particle (grit) sizes for ultrasonic machining. The criteria for selection of an abrasive for a particular application include hardness, usable life, cost, and particle size.

In order of hardness, boron carbide, silicon carbide, and aluminum oxide are the most commonly used abrasives. The abrasive used for an application should be harder than the material being machined; otherwise, the usable lifetime of the abrasive will be substantially shortened. Boron carbide is selected when machining the hardest workpiece materials or when the highest material removal rates are desired. Although the cost is five to ten times greater than the next hardest abrasive, silicon carbide, the usable life of boron carbide is 200 machine-operating hours before cutting effectiveness is lost and disposal is necessary. This compares with a usable lifetime of approximately 60 hr for silicon carbide. The combination of high removal rates and extended lifetime justify the higher cost of boron carbide.

The size of the abrasive particles

械螺钉连接方法所带来的疲劳问题。

7.2.5 磨料

几种磨料有不同粒度可用于超声波加工。选择用于特定应用的磨料标准包括硬度、使用寿命、成本和粒径。

在硬度方面，碳化硼、碳化硅、氧化铝是最常用的磨料。应用的磨料比加工的材料要更坚硬，否则，磨料的可用寿命将大大缩短。在加工最坚硬的工件材料时，或要求最高的材料去除率时选择碳化硼。虽然成本是较次之的磨料碳化硅的5至10倍，但碳化硼的使用寿命为200个机器工作小时，之后才会失去切削效率，必须进行处理。与此相比，碳化硅的使用寿命大约为60个机器工作小时。高去除率和长寿命的组合证明碳化硼的成本较高。

研磨颗粒的尺寸影响所获

influences the removal rate and surface finish obtained. Abrasives for USM are generally available in grit sizes ranging from 240 to 800. While the coarser grits exhibit the highest removal rates, they also result in the roughest surface finishes and are therefore used only for roughing operations. Conversely, 800-grit abrasives will result in fine surface finishes, but at a drastic reduction in the removal rate.

得的表面光洁度。用于 USM 的磨料粒径范围通常为 240~800，而较粗的磨粒去除速率最高，但也得到了最粗糙的表面光洁度，因此只用于粗加工操作。相反，800 粒度的磨料将会产生精细的表面光洁度，但会大大降低去除率。

7.3　Applications of Ultrasonic Machining 超声波加工的应用

Most successful ultrasonic machining applications involve drilling holes or machining cavities in nonconductive ceramic materials. The selection of USM as the preferred manufacturing process usually occurs because no other process is capable of performing the task, the alternative process would require substantially longer process times, or because of a reduction in scrap rates in fragile workpieces.

大多数成功的超声波加工应用包括在非导电陶瓷材料中钻孔或加工腔体。通常选择 USM 为首选的制造工艺是因为没有其他工艺能够承担其任务，替代工艺将需要相当长的加工时间，或因为易碎工件的报废率降低。

As reported by Rutan (1984), special USM tools are often used to simultaneously produce a multitude of holes in precise patterns. This technique significantly increases productivity without compromising quality.

正如 Rutan（1984）所报道的那样，特殊的 USM 工具通常被用来同时加工出许多图形精确的孔洞。这种技术显著提高了生产力而不影响质量。

(1)Used for machining hard and brittle metallic alloys, semiconductors, glass, ceramics, carbides etc.

(2)Used for machining round, square, irregular shaped holes and surface impressions.

(3)Machining, wire drawing, punching or small blanking dies.

（1）用于加工硬脆性金属合金、半导体、玻璃、陶瓷、碳化物等。

（2）用于加工圆形、方形、不规则形状的孔和表面印痕。

（3）机械加工、拉丝、冲孔或小冲裁模。

8 Others non-traditional Machining
其他种类的特种加工

8.1 Plasma Cutting 等离子加工

Plasma has been used for thermal cutting for many years. The heat created by plasma cutting machines is nine times hotter than the surface of the sun. The light from stars results from plasma gases.

Plasma cutting systems use a pressurized conductive gas to cut electrically conductive materials. The difference between regular plasma and precision plasma, also call high-definition plasma and fine plasma cutting systems, is that the precision system cuts more accurately due to increased energy density. See Figure 8-1.

8.1.1 How does Plasma Cutting Work

The common concept of matter is that it has three states: solid, liquid, and gas. But there is a fourth state: plasma. For example:

等离子加工已用于热加工许多年。等离子加工产生的温度比太阳表面的温度还热9倍。加工时的闪烁是由等离子气体产生的。

等离子加工系统使用加压的导电气体来切割导电材料。传统等离子加工和精确等离子加工（又称高精确等离子加工、精细等离子加工）系统的区别在于精确等离子加工系统由于能提供更高的能量，所以可以加工出更高精确度的工件，如图8-1所示。

8.1.1 等离子加工的原理

通常，物体的状态有三种：固态、液态、气态。但是物体还有第四种状态：等离子态。例

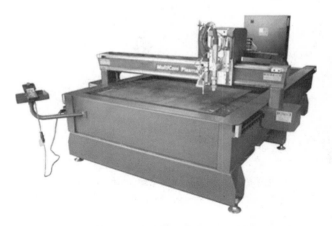

Figure 8-1　Plasma cutting machine
等离子加工机床

the element water can be ice, liquid, steam, or plasma. The amount of heat energy applied to each state changes its qualities.

　　Adding energy to ice results in water. Adding more energy to water creates steam consisting of hydrogen and oxygen. Adding additional energy to steam produces ionized gas, which becomes electrically conductive. This electrically conductive ionized gas is called a plasma. A lightning bolt is an example of plasma. See Figure 8-2.

　　The interior of a plasma cutting torch consists of a consumable nozzle, an electrode, and either gas or water to provide cooling. When the gas flow stabilizes, the power supply provides a high-frequency current to the electrode. This current creates electric arcs inside the nozzle. The flowing gas passes

如：水元素可以是冰、水流、水蒸气或等离子。每次状态转换都会带来能量上的变化。

　　冰变成水要吸热。若是吸收更多的能量，水将变成由氢和氧组成的水蒸气。水蒸气若是继续吸热则会变成导电的电离气体，此种电离气体就称作等离子气体。一道闪电就是一种等离子，如图 8-2 所示。

　　等离子切割器的内部由一个消耗性喷嘴、一个电极、冷却气体或冷却液组成。当气流稳定下来之后电源将向电极提供高频电脉冲。电脉冲将在喷嘴内部产生电弧，气体经过电弧之后将被电离，产生导电性等离子弧。高压

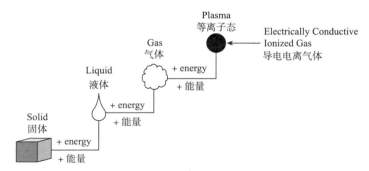

Figure 8-2　The four states of matter
物质的四种形态

through these arcs, becomes ionized, and results in an electrically conductive plasma arc. Pressurized gas flowing through the nozzle then creates a pilot arc as shown in Figure 8-3.

气体从喷嘴喷出，产生如图 8-3 所示的导引电弧。

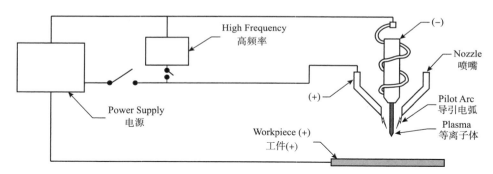

Figure 8-3　High frequency circuit ionizes the plasma and creates a pilot arc.
高频电路产生等离子和导引电弧

As the nozzle approaches the workpiece, the negatively-charged pilot arc is attracted to the positively-charged workpiece. An electronically-controlled sensor switches off the high frequency for the pilot arc. Gas ionization is then sustained from the main DC power supply.

当喷嘴接近工件表面时，电负性的导引电弧被吸引到正电性的工件表面，电控传感器将关闭高频电弧。电离气体主要由直流电源维持。

As the heated plasma arc strikes the workpiece, the plasma arc melts the metal. Then the high pressure gas flow forces out the molten metal and pierces the workpiece. After the piercing occurs, the motion machine is activated and the metal is cut according to the programmed shape.

8.1.2 Plasma Processes

(1) Conventional Plasma Cutting

Conventional plasma generally uses either air or nitrogen to cool and create plasma. This system is used primarily for hand held operations.

(2) Dual Gas Plasma Cutting

Dual gas plasma uses one gas for creating plasma for cutting; the other gas is used to shield the cutting edge from the atmosphere. Various gas combinations are used to create the desired results. See Figure 8-4.

(3) Water Shield Plasma Cutting

Water shield plasma uses a similar process as dual gas plasma cutting. Instead of using a gas for shielding, it uses water to achieve better cooling for the nozzle and workpiece. This is especially useful for cutting stainless steel.

(4) Water Injection Plasma Cutting

In water injection plasma cutting, a single gas is used. Water is injected into the

当等离子热流冲击工件表面时，等离子弧熔化金属，然后高压气体排除熔融金属并穿透工件形成切口。随之，机床开始移动，按编程的形状切割材料。

8.1.2 等离子加工的分类

（1）传统等离子加工

传统等离子加工通常使用空气或者氮气进行冷却并制造等离子体，这种系统通常用于手持加工。

（2）双气流等离子切割

双气流等离子切割使用一种气体切割，另一种气体用于将切好的边缘与空气隔绝。欲达到此种效果，我们可以使用各种各样的气体组合，如图8-4所示。

（3）水盾等离子加工

水盾等离子加工是类似双气流等离子切割的过程，不同的是水盾等离子加工使用水来更好地冷却喷嘴和工件，这对切削不锈钢尤为重要。

（4）水射流等离子加工

在水射流等离子加工中，我们只使用一种气体。水被注射入

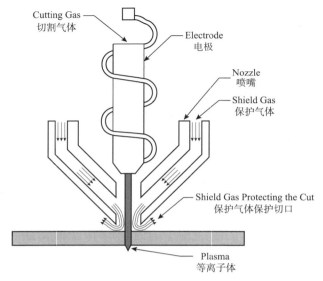

Figure 8-4　Dual gas plasma cutting
双气流等离子切割

plasma arc and increases the energy density of the gas as it leaves the nozzle. This process produces better quality cutting for many materials.

8.1.3　Materials for Plasma Cutting

Plasma cutting requires electrically conductive material. Lasers, in contrast, because they use highly focused light, can cut both electrical conductive and non-conductive materials.

8.1.4　Accuracy

The repeatability of the X-Y-Z movements and the vibrational stability of the machine are critical for cutting accuracy. The distance of the torch to the workpiece also affects part accuracy. If the torch is too high, it can create

等离子弧中，增加了气体离开喷嘴时候的能量密度。这个工艺能使许多材料的切割加工质量大大提升。

8.1.3　可使用等离子加工的材料

等离子加工必须用在导电材料的加工中。而激光加工使用高密集度的光线，可切割导电材料和不导电材料。

8.1.4　精度

X-Y-Z 三轴往复移动或机器的振动稳定性，对加工的精度至关重要，工具到工件的距离也影响零件加工的精度。如果相距太远，电弧将会拉得过长，这样热

a long arc and put excessive heat into the material and cause it to have a beveled edge.

Precision plasma machines have much greater accuracy than conventional plasma machines. Laser cutting machines are still more accurate, but the gap is narrowing between lasers and precision plasma machines.

The general tolerances for production cutting with precision plasma is ± 0.254mm. A single part can be cut within ± 0.1mm. The angularity of a cut on 9.52mm is ± 1 to 3 degrees; below 4.76 mm is ± 0 to 1.5 degrees. Accuracy is also determined by the speed of the cutting machine.

8.1.5 Applications for Plasma cutting

(1) Application of covers with plasmas beam, for example: high quality metal and ceramic covers with thickness 0.1mm which are steady against corrosion, temperature and wear out.

(2) Welding, which is covering the wide area of plasma application.

(3) Cutting and dividing the materials, for example: cutting the thin tins, cutting the aluminum plates, cutting the highest quality steels with thickness to 25mm.

(4) Turning with plasmas beam as of the local heating source of material, before cut of chip.

量发散的范围就更大，将会导致加工边缘变得倾斜。

精确等离子加工机床比传统等离子加工机床的精度更高。激光加工的精度依然更高，但是精确等离子加工和激光加工的差距却正在缩小。

精确等离子加工的公差通常为 ± 0.254mm，单件的精度可以达到 ± 0.1mm，9.52mm 切口的角度公差为 ± 1°~3°，4.76 mm 切口可以达到 0~1.5° 的角。精度通常是由加工速度决定的。

8.1.5 等离子加工的应用

（1）使用等离子给工件镀层，例如：给工件镀上一层厚度为 0.1mm 的高质量金属或陶瓷以增强工件的耐腐蚀性、耐热冲击性和耐磨性。

（2）焊接：使用等离子束将工件焊接成型。

（3）切割材料，例如：切割薄的马头铁、切割铝制板料、切割厚度可达 25mm 的优质钢材。

（4）用等离子束给工件预热。

8.2 Water jet and Abrasive Waterjet Machining 水射流和磨粒流加工

Manufacturers, in their quest for productivity and quality, are searching for alternate ways to cut various materials. Increasingly, they are turning to waterjet and abrasive waterjet machining.

Often the conventional machining method is to either saw cut or plasma cut the material, then machine the finish part. Abrasive waterjet machining often eliminates this secondary operation. Extremely powerful abrasive waterjets are capable of cutting through 3-inch tool steel at a rate of 1.5 inches a minute, 10-inch reinforced concrete at over 1 inches a minute, and 1-inch plate glass at 20 inches a minute.

8.2.1 Fundamentals of Waterjet and Abrasive Waterjet Machining

Modern use of waterjet machining technology dates to 1968. Dr. Norman C. Franz, Professor of Forestry of the University of British Columbia, was issued a patent for developing a high-pressure waterjet cutting system. See Figure 8-5 and Figure 8-6.

A modern waterjet system receives regular low-pressure water. To prevent

生产商通常追求质量和生产率，所以他们一直在寻求能加工更多种类材料的加工方法。如今，越来越多转向于使用水射流或是磨粒流加工。

传统式的加工方法是锯切和等离子加工配合使用，然后加工成型。磨粒流加工通常不进行第二步操作。高能量磨粒流有能力以 1.5in/min 的速度切割 3in 厚的工具钢，或以 1in/min 的速度切割 10in 的加固混凝土，或以 20in/min 的速度加工 1in 的平板玻璃。

8.2.1 水射流和磨粒流加工基础

现代磨粒流加工技术可追溯至 1968 年，是由英国哥伦比亚大学的林业学教授 Norman C. Franz 博士发明的，Norman C. Franz 博士发明了高压水射流加工系统的原型，如图 8-5、图 8-6 所示。

现代水射流加工通常接收低压水流。为了防止增压器泵密封

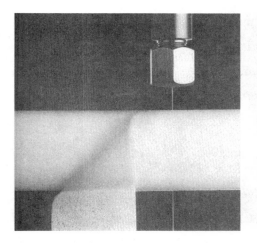
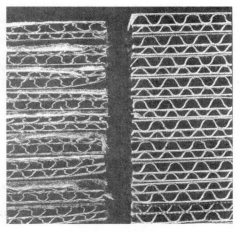

Figure 8-5　Cutting with high-pressure water
高压水射流加工

Figure 8-6　Comparison of cutting corrugated boxboard: With mechanical knife (Left), with waterjet (Right)
加工波纹状盒子的质量对比：
机械切削（左），水射流加工（右）

premature failure of intensifier pump seals, system check valves, and the nozzle orifice, the water is filtered by one, two, or sometimes three series of filter banks. Water free from particles and minerals increases the life of the internal mechanisms of the waterjet system.

This filtered water proceeds through a high-pressure intensifier. The intensifier can increase the pressure from 20,000 to 60,000psi, and accelerate the water to two to three times the speed of sound.

The pressurized water is transferred through high-pressure tubing to a cutting head. The high-pressure water passes through

件、系统止回阀和喷嘴孔过早失效，水流将经过1~3个过滤器组，无杂质的水流能增加水射流机床的寿命。

过滤的水流将经过一个增压器，增压器可以将水压由20,000psi 增至 60,000 psi，并将水流的速度提升至2~3倍的声速。

高压水流将通过高压软管，经过切割头上直径为 0.08~0.6 mm 的小孔被射向加工起始端。当水

a small orifice, 0.08 to 0.6mm, in the cutting head. After the water pierces the material, the used water is collected into a tank or a compact traveling catcher unit. The material is cut when the water pressure exceeds the material's compressive strength.

8.2.2 Materials Cut With Waterjet

Waterjet apart from using any abrasives, can cut many materials. Even thin metals have been cut with high water pressure. Listed are some materials that can be cut with waterjet in Table 8-1.

流穿透材料后，使用过的水流将被收集到一个水箱或一个移动收集装置中。当水压超过材料的抗压强度时，材料将被切削成型。

8.2.2 可以用水射流加工的材料

不使用任何磨粒的水射流加工方法可以加工任意种类的材料，即使是薄钢板也可以用高压水射流加工。表 8-1 中列出了一些可用水射流加工的材料。

Table 8-1 Materials that can be cutted with waterjet 可以用水射流加工的材料

Thin Aluminum 薄铝板	Lead 铅	Fiberglass 玻璃纤维
Plexiglass 钢化玻璃	Kevlar 纤维	Circuit Boards 电路板
Corrugated Boxboard 蜂窝板	Wood 木材	Rubber 橡胶
Graphite Composite 石墨复合物	Carpet 地毯	Food Products 食品

8.2.3 Accuracy

High-accuracy positioning devices can cut within ±0.127mm and closer. Production cutting uses a tolerance of ±0.127 to 0.38mm. Stock thickness greatly affects cut accuracy.

The top surface of thick cuts tend to be smoother than the bottom, and the bottom cuts tend to be tapered and have jet-induced

8.2.3 精度

高精度加工设备可以达到 ±0.127mm 或更高的精度。加工通常使用的公差数值为 ±0.38mm。材料的厚度会显著影响加工精度。

顶部的加工表面通常比底部要光滑，切削后的底部通常是锥形的，并具有喷射引起的条纹。

striations. Slowing down the cutting speed improves the surface finish.

8.2.4 Safety

Waterjet can cut flesh and bone easily; therefore, operators must be protected from the high-pressure forces. Likewise the noise level on some machines, from 80 to 120DB, poses a hazard and require ear protection. To lower noise level, some machines cut under water which reduces the noise level to under 75DB.

8.2.5 Advantages of Waterjet and Abrasive Waterjet Machining

(1) Material Savings

Much material waste often occurs with sawing and milling. Since abrasive waterjet cutting produces a small kerf, parts can be nested for maximum material utilization.

(2) No Special Tooling Required

Waterjet permits parts to be cut without special tooling or cutters. The desired shape is programmed and downloaded into the waterjet system. When a design needs to be changed, the program can be altered with little production downtime.

Also, when material or workpiece thickness changes, only the cutting speed parameters need to be changed. This provides great flexibility and permits just-in-time machining.

8.2.4 安全

水射流可以轻易地切割骨肉，所以，操作人员应该远离高压装置。有些机器上的噪声会达到80~120dB，这是对人体有害的，所以需要耳部保护。为了降低噪声，有些机床在水下切割，这样能将噪音降至75dB。

8.2.5 水射流和磨粒流加工的优势

（1）节省材料

锯和磨通常会耗费更多材料。由于磨粒流造成的切口小，零件可以嵌套以最大限度地利用材料。

（2）不需要特殊刀具

水射流加工不需要特殊的刀具。设计好的形状可以直接导入加工系统中进行加工。当需要更改设计时，可以很快对程序进行改动。

类似的，当材料或工件的厚度发生改动时，只需要改动加工速度即可。这就保证了加工的灵活性和实时性。

(3) Moisture Absorption Not a Problem

Materials such as corrugated board, paper, and carpet absorb little moisture because of the high velocity of the waterjet system. Any dampness that may occur rarely affects the product.

(4) Focusing the Waterjet is Not Critical

Unlike lasers, where focus is critical, the standoff distances of waterjet, generally under one inch (25mm), produce little cut variation. However, optimum distance is usually twice the diameter of the mixing tube's ID.

(5) Simple Fixtures Required

Since the waterjet cutting forces are extremely low, from 3 to 5 lbs (1.36 to 2.27kg), simple fixtures can hold parts. However, materials with internal stresses must be secured to prevent the parts from moving while being cut.

(6) Entry Hole Not Required

Abrasive waterjet produces its own entry holes. Parts with multiple cavities can be machined without any other premachining.

(7) No Heat Affected Zone or Microcracks

Waterjet and abrasive waterjet machining create no heat-affected zone at the cut edge, as do EDM, lasers or plasma cutting

（3）材料的吸水现象不会影响加工

类似蜂窝箱、纸、地毯等在加工时由于水射流系统的高速而吸收的水分很少，这对于产品的影响微乎其微。

（4）水射流的焦距要求不是十分严格

与激光加工的严格焦距不同，水射流喷嘴距工件表面的距离不是十分严格，通常在1in以下，这样加工出的工件不会有太大误差。然而，焦距通常取喷嘴直径的2倍。

（5）简单的加持装置

由于水射流的切削力不是太大，通常为1.36~2.27 kg，所以简单的夹持工具就可以夹持工件。然而，必须固定具有内应力的材料以保证在加工的过程中不会发生移动。

（6）不要求加工起始孔

水射流加工可以自己加工出起始孔，具有很多孔腔的工件不需要任何前期处理就可以加工。

（7）无热影响区和微裂缝

水射流和磨粒流加工在切削边缘处没有热影响区，而电火花加工、激光加工、等离子加工却

machines.

The cold-cutting operation protects the microstructure of the cut surfaces, and prevents edge work hardening, microcracks and heat distortion. One company discovered that their titanium dental implants cut with lasers suffered from microscopic heat-stress cracks. These cracks permitted harmful bacteria to develop in the implants. Abrasive waterjet solved the problem.

(8) No Fumes

Lasers require a system to remove cutting fumes. Neither waterjet or abrasive waterjet cutting produces fumes.

(9) Eliminates Some Difficult Cutting Problems

To drill through carpet is an extremely difficult operation because the carpet tends to wrap around the drill. Waterjet machining eliminates this problem. Laminated products can also be cut without separating the laminations.

Traditional cutting of composites often damaged them with heat, fraying or edge delaminating. In addition, diamond or carbide-tipped routers, bandsaws or abrasive wheels cut slowly. Abrasive waterjet solves these composite cutting problems.

有热影响区。

这种冷切削过程保护了被加工表面的微观结构，并阻止了边缘处的硬化、微观裂纹和热变形。某企业发现其使用激光加工做的钛合金牙齿产生了热应力裂纹，这种裂纹会导致有害细菌在裂缝中滋生，水射流加工解决了这一问题。

（8）无烟尘产生

激光加工需要一单独系统来清除切割烟尘，水射流加工和磨粒流加工都解决了这一问题。

（9）解决了某些加工困难的问题

在地毯上打孔是一件很困难的事情，因为地毯在加工过程中会弯曲。水射流加工解决了这一问题，层压产品也可以在不分层的情况下加工。

传统对复合物加工的方式通常会由于热、磨损或边缘分层现象而损坏材料。另外，在钻石或硬质合金刀具、带锯或砂轮的加工中要减慢速度。磨粒流解决了这一问题。

8.2.6 Disadvantages of Waterjet and Abrasive Waterjet Machining

(1) Frosting From Abrasive Waterjet Cutting

During some applications frosting can occur from abrasive waterjet cutting. When frosting is unacceptable, a dehazer device can be attached to the cutting head. With this device low-pressure water surrounds the cutting stream and prevents the high-pressure abrasive stream from frosting the material. The dehazer is particularly useful when cutting glass, brass, marble or other shiny materials. Its added benefit is that it lowers the noise level and the airborne abrasive dust. It also increases the standoff with no appreciable change in kerf.

(2) Slower Speed Rates and Higher Costs Than Plasma and Lasers

Plasma cutting machines and lasers cut faster and cheaper, but both leave a heat affected zone. However, thick materials limit laser cutting ability.

(3) Catchers Needed With Multi-Axis Cutting

Because of the high-velocity abrasives leaving the cut, a catcher system needs to be behind the cutting head to prevent part damage. Lasers generally do not have this

8.2.6 水射流和磨粒流加工的缺点

（1）磨粒流加工的结霜现象

在磨粒流加工过程中可能会发生结霜现象。当结霜现象不可接受时，可以在加工起始处加上除霜器。在加工处周围的低压水阻止了高压水的结霜现象。在切削玻璃、铜、大理石和其他光亮的材料时，除霜器通常是十分重要的。

（2）速度慢，花费大

等离子加工和激光加工的加工速度更快，而且更加便宜，但是这两种加工方式都有热影响区。然而，厚工件限制了激光加工的切削能力。

（3）多轴联动加工，需要反馈监测系统

由于磨粒的速度极高，在加工的起始点应加上反馈监测系统来防止工件损坏。激光加工通常不需要这类系统。

Unit 8 Others non-traditional Machining 其他种类的特种加工

problem.

(4) Large Cuts Become Stratified

Thick cuts tend to produce stratified surfaces. The waviness of the cut is greatest on the bottom of the cut. In contrast, the electrode of wire EDM travels along the entire surface of the cut and produces an exceptionally smooth surface, even for the first cut.

（4）厚工件的加工将会导致分层现象

加工厚的工件时会导致表面分层。在切削的底部，摆动现象更加明显。相反的，电火花加工的电极加工时直接经过整个表面并会加工出更加光滑的表面，即使是初加工也是如此。

8.3 Electron Beam Machining 电子束加工

In electron-beam machining (EBM), electrons are accelerated to a velocity nearly three-fourths that of light (~200,000km/sec). The process is performed in a vacuum chamber to reduce the scattering of electrons by gas molecules in the atmosphere. The electron beam is aimed using magnets to deflect the stream of electrons and is focused using an electromagnetic lens. The stream of electrons is directed against a precisely limited area of the workpiece; on impact, the kinetic energy of the electrons is converted into thermal energy that melts and vaporizes the material to be removed, forming holes or cuts.

电子束加工中，电子的速度可以接近四分之三的光速（近乎200000 km/s）。这个过程应在真空装置中进行，以减少电子在空气中出现的发散现象。电子束加工使用磁场来偏转电子并使用电磁透镜来达到这一目的。电子束直接射在工件的一个小范围表面上，此时，电子的动能将会转化为热能并熔化蒸发被移除的材料，产生孔或对工件进行切割。

Typical applications of EBM are annealing, welding, and metal removal. A hole in a sheet 1.25mm thick up to 125um diameter

电子束加工的典型应用是退火、焊接、材料移除。电子束加工可以直接加工1.25 mm厚、直

can be cut almost instantly with a taper of 2 to 4 degrees. EBM equipment is commonly used by the electronics industry to aid in the etching of circuits in microprocessors.

径 125 μm、有 2°~3° 锥角的孔。电子束加工装置通常应用在电子束加工工厂中来蚀刻电路板。

8.4　Ion Beam Machining　离子束加工

In simple terms ion beam machining can be viewed as an atomic sand blaster. The grains of sand are actually submicron ion particles accelerated to bombard the surface of the work mounted on a rotating table inside a vacuum chamber. The work is typically a wafer, substrate or element that requires material removal by atomic sandblasting or dry etching.

A selectively applied protectant, photo sensitive resist, is applied to the work element prior to introduction into the ion miller. The resist protects the underlying material during the etching process which may be up to eight hours or longer, depending upon the amount to be removed and the etch rate of the materials. Everything that is exposed to the collimated ion beam (may be 15″ in diameter in some equipment) etches during the process cycle, even the resist.

In most micromachining applications the desired material to be removed etches

离子束加工可以被看作是使用原子砂轮进行磨削。磨粒实际上就是加速撞击工件表面的离子束，该工件通常被放置在真空装置内的工作台上。工件通常是一片干胶片、底片以及需要进行微观材料移除或是干燥蚀刻的工件。

在离子束加工之前，我们使用光敏选择性保护装置。这种保护装置在冲蚀过程中保护了工件，使得冲蚀过程将会长达 8h 或更长，时间的长短取决于材料的电蚀速度。在加工过程中被离子束（在某些设备中直径可能为 15 英寸）直接照射到的物体都会被冲蚀，即使是保护装置。

在大多数微加工中，材料被切除的速度为被保护的加工过程

at a rate 3 to 10 times faster than the resist protectant thus preserving the material and features underneath the resist.

Ion beam milling is used in fabricating electronic and mechanical elements for a wide variety of commercial, industrial, military and satellite applications including custom film circuits for RF and microwave circuits.

的 3~10 倍。

离子束加工用于制造商业、工业、军事、人造卫星等的电子或机械元件，甚至电影底片，或加工微细电路。

Questions 习题

1. How does plasma cutting work?
2. What kind of gas is used for plasma cutting?
3. What are the four states of matter?
4. What kinds of materials can plasma cut?
5. How does waterjet cutting work?
6. What type of materials is generally used for waterjet cutting?
7. List the advantages of waterjet and abrasive waterjet machining.
8. Describe the disadvantages of waterjet and abrasive waterjet machining.

1. 等离子加工是如何工作的？
2. 等离子加工使用的气体是什么类型？
3. 什么是物质的四种形态？
4. 等离子加工可以加工什么种类的材料？
5. 超声加工是如何工作的？
6. 超声加工可加工什么类型的材料？
7. 简述水射流加工的工作原理。
8. 水射流加工可加工什么类型的材料？

References

[1] Hassan El-Hofy. Advanced Machining Processes: Nontraditional and Hybrid Machining Processes [M]. 2005.

[2] Liu J.C., Zhao W. Q. Zhao W. S.. Non-Traditional Machining [M]. 2000.

[3] Carl Sommer. Non-Traditional Machining Handbook [M]. 2000.

[4] Jackson M. J.. International Journal of Manufacturing Technology and Management [M]. 2005.

[5] Xia F. F. Research on the Ni-TiN Nanocomposite Coating Prepared by Ultrasonic-Electrodeposition [D]. 2008.

附录　Words and Expressions

dielectric oil 绝缘油液

ionize 电离

electrical impulses 电流脉冲

vaporize 气化

melt 溶解

pressurized 被压缩的

suspended 悬浮的，暂停的，缓期的

particle 颗粒　质点

chiller 冷却装置

machining accuracy 加工精度

a DC servo mechanism 直流伺服机械装置

positive 阳性的

negative 阴性的

insulator 绝缘体

amps 安培

electrode 电极

insulating 绝缘的　隔热的

plasma channel 等离子通道

polarity 极性

carbide 碳化物，碳化合金

titanium 钛

copper 铜

graphite 石墨

plate 电镀

nodule 小节，小瘤

fume 烟气

ventilation 通风设备

boron 硼

beryllium 铍

toxic 有毒的

erode 侵蚀

flush 冲洗

manufacturer 制造商，厂商

coolant 冷却液，冷冻剂，散热剂

flash point 引火点

precaution 预防措施

infrared scanner 红外线扫描仪

cavity 腔，洞

erratic 不稳定的

arc 形成电弧

unwanted 不需要的，有害的，讨厌的

roughing operations 粗加工

finishing operations 精加工

suction 抽吸，吸力

pressure gauge 压力表

stud 螺栓

spike 长钉

grinder 磨床

needle nose pliers 尖嘴钳

fuzzy logic 模糊逻辑

seal 密封

secondary discharge 二次放电

convex 凸面的，凸状体

jet 喷射，射出

nozzle 喷嘴，管口

square hole 方孔

Abrasive flow machining 磨料流加工

ultrasonic a. 超声（波）的

　　　　　n. 超声波

ultrasonic generator 超声波发生器

slurry n. 稀（泥，砂）浆，悬浮体［液］

transducer n，换能器，变频器

turbulent a. 湍［紊］流的

hypodermic needle 皮下注射针头

piezoelectric a. 压电的

magnetostrictive a. 磁致伸缩的

lead zirconate titanate 锆钛酸铅

acoustic a. 听觉的，声［学］的

Monel 蒙乃尔高强度耐蚀镍铜合金

music wire 琴弦

grit size 磨料粒度

tuning adjustment 调谐匹配

pantograph n. 缩放仪

pantograph profile grinder 缩放仪式轮廓磨床

pyroceram n. 耐高温陶瓷黏合剂，耐热［耐高温］玻璃